Environmental Justice and Farm Labor

Utilizing a model derived from literature on environmental justice overlaid with multiple scales of agriculture, *Environmental Justice and Farm Labor* provides key insights about laborers in agriculture in the United States. It addresses three main topics: (1) justice-related issues facing farmers and laborers on farms; (2) how history and policy have impacted them; and (3) the opportunities and leverage points for change in improving justice outcomes.

It explores who labors in US agriculture and the justice-related issues facing these workers, including occupational injury and illness, lack of access to health-care, substandard housing, hunger, low wages, issues pertaining to immigration, and the inability to organize. In addition, it assesses the impacts of labor safety, immigration and international policy, and in particular the effects of organic and fair trade certification. Two detailed case studies, one based on conventional agriculture in Florida and the other on organic agriculture in the Northeast, highlight the interrelated but unique challenges facing those who labor in the different sectors of this complex agricultural system.

Finally, it touches on justice claims and the role of grassroots activism in improving justice outcomes by highlighting organizations operating at multiple scales to contribute to the livelihood of farmers and laborers in the different areas of agriculture.

Rebecca E. Berkey is Director of Service-Learning, Northeastern University, Boston, USA, and Associate Director, Northeastern Environmental Justice Research Collaborative, USA.

Earthscan Food and Agriculture

The Sociology of Food and Agriculture
Second edition
Michael Carolan

Food Security, Gender and Resilience
Improving smallholder and subsistence farming
Edited by Leigh Brownhill, Esther M. Njuguna, Kimberly L. Bothi, Bernard Pelletier, Lutta W. Muhammad and Gordon M. Hickey

Climate Change and Agricultural Development
Improving resilience through climate smart agriculture, agroecology and conservation
Edited by Udaya Sekhar Nagothu

Forgotten Agricultural Heritage
Reconnecting food systems and sustainable development
Parviz Koohafkan and Miguel A. Altieri

International Law and Agroecological Husbandry
Building legal foundations for a new agriculture
John W. Head

Food Production and Nature Conservation
Conflicts and Solutions
Edited by Iain J. Gordon, Herbert H.T. Prins and Geoff R. Squire

Sustainable Intensification in Smallholder Agriculture
An integrated systems research approach
Edited by Ingrid Öborn, Kwesi Atta-Krah, Michael Phillips, Richard Thomas, Bernard Vanlauwe and Willemien Brooijmans

Environmental Justice and Farm Labor
Rebecca E. Berkey

Plantation Crops, Plunder and Power
Evolution and exploitation
James F. Hancock

Food Security, Agricultural Policies and Economic Growth
Long-term dynamics in the past, present and future
Niek Koning

For further details please visit the series page on the Routledge website:
http://www.routledge.com/books/series/ECEFA/

Environmental Justice and Farm Labor

Rebecca E. Berkey

Routledge
Taylor & Francis Group
LONDON AND NEW YORK
earthscan
from Routledge

First published 2017 by Routledge

2 Park Square, Milton Park, Abingdon, Oxfordshire OX14 4RN
52 Vanderbilt Avenue, New York, NY 10017

Routledge is an imprint of the Taylor & Francis Group, an informa business

First issued in paperback 2019

British Library Cataloguing in Publication Data
A catalogue record for this book is available from the British Library

Library of Congress Cataloging in Publication Data
Names: Berkey, Rebecca E., author.
Title: Environmental justice and farm labor / Rebecca E. Berkey.
Description: London ; New York : Routledge, 2017. | Series: Earthscan food and agriculture series | Includes bibliographical references and index.
Identifiers: LCCN 2016039035 | ISBN 9781138183155 (hbk) | ISBN 9781315645995 (ebk)
Subjects: LCSH: Environmental justice–United States. | Agricultural laborers–United States–Social conditions. | Farmers–United States–Social conditions. | Agriculture and state–United States. | Agricultural industries–Environmental aspects–United States.
Classification: LCC GE230 .B47 2017 | DDC 331.7/630973–dc23
LC record available at https://lccn.loc.gov/2016039035

ISBN: 978-1-138-18315-5 (hbk)
ISBN: 978-0-367-22779-1 (pbk)

Typeset in Bembo
by Taylor & Francis Books

This book is dedicated to the millions of hands that feed us—providing fresh, nutritious food to our tables and homes. And, to the giants upon whose shoulders I stand, thank you for paving the way to challenge, learning, and discovery.

Contents

Illustrations

Figures

Tables

Acknowledgements

I have many people to thank for their guidance, encouragement, challenge and support as represented in the pages of this book. First, to my colleagues and collaborators that are champions of farmworker advocacy and justice—Jeannie Economos, Holly Baker, and Elizabeth Henderson—your advice, guidance, and perspective has been invaluable in shaping my thinking around and exposure to issues facing farmworkers in both conventional and organic agriculture. In addition, I am so appreciative of the feedback each of you offered throughout various stages of this project. I have learned from each of you the value of tenacity in the face of seemingly insurmountable odds, and that remaining dedicated to your passion areas over time can, and does, make change.

To my academic colleagues and confidants, thank you for your edits, advice, and feedback. Daniel Faber, Christopher Bosso, and Erika Koss, I am incredibly grateful for the insight you contributed to this project—from nudging me to find and include my own voice in the work, to acknowledging and seeking to understand my reverence and respect for those who labor on our farms, to your kind confidence in me. Tania Schusler, Steve Chase, and Joseph Siry, I thank the three of you as well for your guidance and investment throughout my doctoral work and beyond. The sage advice you have selflessly shared with me in the time I have known each of you is reflected in each word, and your mentorship and modeling were continually in my mind as I reflected on how to capture and portray not only the complexity of this topic, but also the real lives of the people this system impacts. In addition, gratitude to my Northeastern University writing group of women and gender non-conforming warrior academics, I deeply admire each of you, and the space we had with one another throughout the year was deeply impactful on my thinking around this project and the path needed to complete it.

Finally, to the rest of my village, thank you for still loving and standing beside me even though this project has split my attention in more directions than I care to admit. I owe a deep debt of gratitude to Emma Clouse, who not only helped to edit and format this book, but also talked me down from many ledges and put up with me when I was being unreasonably insecure or neurotic. I wish you all the best for your Ireland adventures, and you are going to undoubtedly be a force for change in this world. To my team and colleagues at

Northeastern University, thank you for listening to me talk about all of this and always checking in on it. To my friends and family, I am so thankful for your belief in me. To my husband Jonathan and stepson Kaiman, I look forward to continuing to explore these and other difficult, complex issues with the two of you, my family, in whom I place my fullest trust and love.

Introduction

Environmental justice and farm labor

Like many people, I grew up without much of an understanding of how our farm system worked. Born and raised in Indiana, I was surrounded by corn, wheat, and soy fields, all crops that are produced in a highly mechanized fashion. Therefore, I believed that widgets and machines were responsible for gathering the fruits and vegetables that I saw at the supermarket and in restaurants. Imagine my surprise when I moved to Florida, and one day found myself on a long bike ride right outside of Orlando where I passed large citrus groves. I felt as though I had taken a step back in time, or was watching a movie of some sort, as I saw hundreds of black and brown bodies in the hot sun, picking fruit off the trees. Now this may sound naïve, but I do believe my story is not unique. It is well understood that most people do not know that most of the fruits and vegetables they purchase at the store (along with the dairy and meat they purchase, the coffee they drink, the chocolate bars they consume, and so on) come from an industry where human labor is one of the biggest inputs to production. As my curiosity grew and I began peeling back the layers of what I saw on that bike ride through research and discovery, the more I learned about how little the system of agricultural labor has changed in many regions of the country since the days of slavery, particularly in some parts of the sector. The aim of this book is to give a fuller picture of who labors in this system, the inherent injustices they face, as well as how these issues manifest themselves in both different and overlapping ways from small-scale organic to large-scale conventional agriculture. Working from foundational and current literature in environmental justice, this book will cover key insights about the laborers in both conventional and organic agriculture in the United States. In doing so, it will consider the following questions: (1) what are the justice-related issues facing farmers and laborers on farms in the United States?; (2) which policies have most impacted the workforce in American agriculture and what are the historical dimensions of those policies?; and (3) what are the opportunities and leverage points for change in improving justice outcomes for those who work in agriculture?

In a letter to the Environmental Protection Agency (EPA), responding to their draft of the Environmental Justice (EJ) 2020 Action Agenda Framework, Jeannie Economos stated on behalf of the Farmworker Association of Florida that:

> Farmworkers are a special and often overlooked environmental justice community. Because farmworkers comprise a "community" that is identified by their occupation, and, in general, oftentimes, by their mobility and their rural demographics, they are not often included in the prevailing discussions around environmental justice. Farmworkers are a community that spans the entire country wherever they go to perform the important agricultural work that this country – and their livelihoods – depend upon. Yet, farmworkers *are* a community, because of their common experiences, interests, concerns and exposures, most notably, their exposures to pesticides in the workplace and, through drift and/or take home exposures, in their home environments.
>
> (Economos, 2015)

It is the definition of community and the framing put forward in this response that undergird what follows throughout this book, and the notion that using the framework of environmental justice to examine the plight of farmworkers may shed new light on how to make change with the hundreds of thousands of individuals who labor on farms in the United States, who are many times exposed to the very things we are afraid of in our food. I will make the case in what follows that not only are farmworkers an environmental justice community, but that this is the case across the continuum of agriculture for overlapping but varying reasons, from small- and family-scale organic to large-scale conventional. In doing so, I also acknowledge that farmworker issues have been explored in various literature and media outlets throughout the years, but will share the perspective I have gained in studying them – that while they have been examined and it is clear there is a panoply of issues facing this class of workers, nothing much has changed in their overall story – and it could be argued that it has gotten worse. That said, environmental justice and the literatures of social change, coalition building, and asset-based community development offer a fresh look at this subject and a framework through which to consider change. This, in combination with the renewed interest in food and our food system, gives hope that another book of this scope could tell a different story in the future.

Labor across the continuum

An environmental justice lens is useful in exploring the issues encountered by farm laborers across all scales of the agricultural continuum, and also for exploring what can and should be done for and with these workers to create more equitable and desirable livelihoods. While farmworker communities themselves have seen their connection to the environmental justice movement, scholars and activists within the movement itself have not always made the connection as clear as it could be. Figure 0.1 summarizes information covered more fully in Chapter 1 regarding the intersections of environmental justice and farm labor for the purposes of providing a framework to explore the different dimensions of injustice within farm labor across all scales of the continuum.

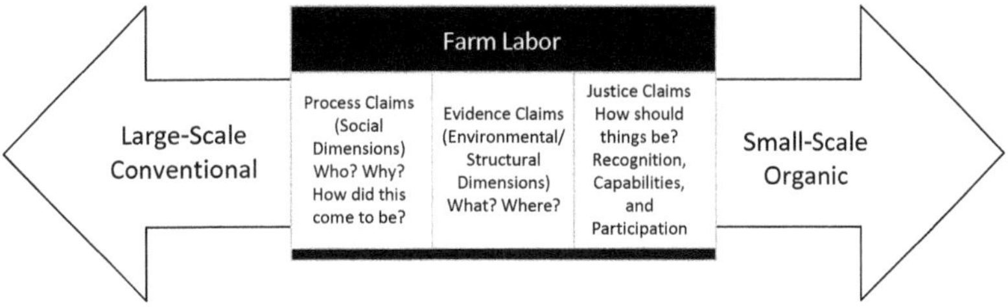

Figure 0.1 Continuum of farm labor and environmental justice claims

Using this model throughout this book, I will consider from multiple angles this complex system, including its historical and systemic roots, the impacts and symptoms of labor-related injustice, as well as promising models for providing justice for laborers at all scales of agriculture.

Structure of the book

Moving through what follows, I provide an overview and in-depth insight into the connections of labor within agriculture in the United States as an environmental justice issue and of farmworkers as an environmental justice community, as well as some of the work being done with and for farmworkers and laborers at various scales within that system. Chapter 1 describes elements of the theoretical and philosophical frameworks of environmental justice, critical theory, community development, and coalition building to create a unique lens through which to examine some of the work being done to, for, and with farmworkers in U.S. agriculture. Specifically, some scholars and activists have called for a more expansive understanding of justice in environmental justice, both in scale and theoretical underpinnings. Farmworkers across the United States face a number of justice-related issues, including labor rights violations, exposure to pesticides, poor working and living conditions, and wage abuse, all of which relate to the core problem for farmworkers in the United States—lacking voice in the policies and decisions that affect them. Given the framing and foundational literature as well as emerging calls for its expansion, farmworkers as a community that is constantly moving, changing, and shifting offer a way to think about, discuss, and see evidence of different types of justice in action.

To delve more deeply into farmworker-specific struggles and the process and evidence claims introduced in Chapter 1, Chapter 2 discusses the infinitely complex system of agriculture within the United States, and the complicated journey that our food takes from the field to our plates. Getting a sense of where farmers and workers fit within this system necessitates first a survey of data about agricultural workers more generally, the historical context within which they are situated, and what we know about how conventional and organic agriculture compared to one another. In this chapter, I will detail the landscape of labor in farming by setting the context about who is working in agriculture in the United States, and then by reviewing key insights from studies on the obstacles and disenfranchisement that keep farm labor marginalized within a broader political-economic system.

Chapter 3 includes a survey of some of the policies that are currently affecting laborers on farms in the United States with the aim of providing the process claims related to the origins of injustice, or broader social dimensions, impacting farmworkers. Agricultural laborers are one of the only classes of workers excluded from the protections and provisions of the Fair Labor Standards Act, which largely stems from the historical roots of farm work in slavery. This has implications today as policies such as the Farm Bill and AgJOBS are debated, decided, and implemented. How and which policies apply to and affect laborers vary between conventional and organic agriculture. In addition, Chapter 3 examines the political-economic historical context of immigrants working in agriculture in the United States and how this has been affected by free trade policy, such as the North American Free Trade Agreement (NAFTA). NAFTA is particularly notable because it is a free trade agreement that made it simpler for agricultural corporations to hire and exploit immigrants as a cheap, unskilled labor force and was passed at a time in U.S. history when conventional agriculture and the mass production of food were becoming increasingly commoditized. The theoretical framework formed in Chapter 1 provides the lens through which I explore the history and impact of policy on farmworkers, and how key events and policies have contributed to a lack of justice for laborers in U.S. agriculture, both organic and conventional.

Chapter 4 describes a richly documented, in-progress case of the Farmworker Association of Florida, specifically in its Apopka, Florida, location. The work of this organization includes worker justice, pesticide actions, immigrants' rights, health education, research projects, and disaster response. The Farmworker Association is a grassroots, community-based organization of over 10,000 member families who work primarily in the vegetable, citrus, mushroom, sod, fern, and foliage industries in 12 counties throughout Central and South Florida. The members are approximately 94 percent Latino (predominately Mexican, Guatemalan, and Salvadoran), 3 percent Haitian, and 3 percent African American, and approximately 40 percent are women. In total, the Farmworker Association of Florida's organizing efforts affect at least 40,000 individuals in rural agricultural areas each year. Each of these populations has its own unique set of needs and issues, ranging from the multiple issues that arise with immigrant and migrant laborers to the long-term health issues of those who have worked on farms in the community for generations. By discussing the work of the Farmworker Association, readers will gain familiarity with the issues facing the two primary groups of workers they serve: (1) immigrants who work as migrant laborers; and (2) members of the African American community who have worked the land there for generations.

In Chapter 5, I will present the findings from my dissertation research as an empirical example of justice issues facing farmers and laborers in organic agriculture. The mixed methods study presented in this chapter builds upon literature and research in environmental justice, public participation, and community development to examine how justice-related issues impact farmers and workers on organic farms in the northeastern United States. It also examines how involvement in a grassroots organization affects how farmers and workers address these issues. At the core of the study is an exploration of environmental justice and its applications at a broad, systemic level; an examination of the current context of laborers in organic agriculture in the Northeast; and finally an investigation of the effects of grassroots organizing in the Northeast Organic Farming Association (NOFA) as it intersects with notions of justice. The first phase of this study comprised an online survey sent out to all NOFA member farmers to gather information about who the laborers on organic farms in the NOFA network are and about the unique justice issues they face. The results of the survey indicate that most of the farms within the network are small-scale and rely heavily on family members and volunteers for labor. Farmer participants identified challenges encountered in enacting their justice-related values, which many times are the reason they choose to farm organically or work on organic farms in the first place. The second phase consisted of phenomenological interviews with farmers and workers from three different member farms known in NOFA for a commitment to justice, to discover patterns of meaning around justice and the contributions of network membership to its realization, as well as interviews with members of their Interstate Council, which offer insight not only into the challenges for organic farms, but also possibilities for change.

Social capital concerns the connections between and among people, and our capacity to use relationships as a resource to accomplish and work toward our goals (Phillips & Pittman, 2009). The implications of these realizations for farmworkers are immense, given the political economic terrain, emphasizing a need to create avenues to build social capital and capacity for strategy enactment. Numerous organizations are working for the betterment of farmworkers in U.S. agriculture. In Chapter 6, I include some of the most pertinent issues facing farmworkers and describe work being done in other areas/regions of the United States, and use that as the framework to introduce some of the different approaches to change in three areas: (1) farmworker associations; (2) coalitions formed around the justice issue of pesticide use and poisoning; and (3) broad-based organizations focused on farmworkers' rights. In addition, this chapter discusses the impact of various fair trade certifications on worker justice.

Finally, the Conclusion draws together various themes from the book, resituating them in the theoretical frameworks of environmental justice, critical theory, community development, and coalition building. In addition, it weaves together the findings of the two case study chapters to provide a context for further exploration and an understanding of the overlapping and distinct concerns within conventional and organic agriculture.

Note to readers

It is my hope that those reading this book—whether academics, students, activists, or farmworkers themselves—will take the time to reflect and think deeply about their own personal relationship to those who labor in our food system. One of the very few things that unites all of us is our need for nourishment, which we gain from food grown across our country and around the world. This food is harvested by people who are undervalued, systemically mistreated and put in positions of subordination and control where they have to tolerate that mistreatment, and whose hands, lives, passions, and fears are a part of each of us. While what follows paints a grim picture, it is also important for us to remember that things will change only if and when we think, reflect deeply, listen, and act, and there are countless examples of people working tirelessly from within and outside of the farmworker community to make that very change.

References

Economos, J. (2015). Re: Draft EJ 2020 Action Agenda Framework. Letter to the Environmental Protection Agency (EPA), July 13.

Phillips, R., & Pittman, R. H. (2009). *An introduction to community development*. London: Routledge.

1 Environmental justice and framing labor in agriculture as a justice issue

A brief history of environmental justice

Definitions and origins

Environmental justice is a movement that focuses on a solution-oriented approach centered on creating a society that is productive and sustainable for all of its members, regardless of race, ethnicity, class, gender, or citizenship status (Pellow, 2000; Sze & London, 2008). As a concept, environmental justice has historically been questioned and debated. Bunyan Bryant's definition of environmental injustice has been lauded as one that represents traditional notions of the field, "the systematic exclusion of people of color from environmental decisions affecting their communities" (1995, p. 5). The environmental justice (EJ) movement expanded the focus of much of the traditional environmental movement, which was solely on the effects on the non-human world such as land, air, animals, etc., to the exclusion of human beings. EJ began in its early days with a focus on environmental racism, or the unequal distribution of environmental benefits and the burden of pollution in communities of color. This evolved into considering broader issues of environmental inequality, expanding to include class, gender, citizenship, and their interconnections (Faber, 2008). Bryant's definition also clarifies the interplay of policies, regulations, and economic context, stating:

> Environmental justice (EJ) … refers to those cultural norms and values, rules, regulations, behaviors, policies, and decisions to support sustainable communities where people can interact with confidence that the environment is safe, nurturing, and productive. Environmental justice is served when people can realize their highest potential.
>
> (Bryant, 1995, p. 6)

Justice is most elusive to those marginalized within a society. For the purposes of this book, I define marginalized peoples as peoples trivialized, devalued, or otherwise excluded from some facets of society based on their cultural, social, and/or historical background(s) (Chambers, 1997; Freire, 2000; Johnson, 2006;

Pellow & Brulle, 2005). Another term used to describe these oppressed segments of the population is subaltern, or people "that are being *selectively victimized* to the greatest extent by corporate environmental abuses" (Faber, 2008, p. 16) and those experiencing "multiple forms of political domination, cultural oppression, economic exploitation, environmental degradation, and resistance" (ibid., p. 238), keeping them from fully expressing their voices in society. I will use these definitions as a background against which to discuss laborers within the different scales of agriculture in the United States as a marginalized population, or part of the subaltern. As the subaltern represents the potential intersectionality of multiple forms of oppression (gender, class, ethnicity, race, etc.), it also must be noted that farmworkers are a population who experience multiple forms of oppression, putting these individuals in a disadvantaged position to live healthy, whole lives and to advocate for their own well-being.

While the current focus of the movement is on the subaltern, most literature states that the environmental justice movement emerged from the antitoxic (1970s and 1980s) and civil rights (1950s and 1960s) movements in the United States. However, historically, the farmworker justice movement predates that of environmental justice and it could be said that the EJ movement draws inspiration from those movements related to farmworkers. In his book, *Forcing the Spring: The Transformation of the American Environmental Movement*, Robert Gottlieb explains how concerns shifted within the antitoxic movement to more community-based models of participation and social change, how groups of people and organizations began to explore their place as a movement overlapping with, but with distinct concerns from, sole consideration of environmental activities and agendas (Gottlieb, 1993). Further, Daniel Faber (2008) characterizes environmental justice as the convergence of several social movements: civil rights, occupational health and safety, indigenous lands, immigrant rights, and environmental health movements, in addition to the community-based movements for social and economic justice, human rights, peace, and solidarity movements.

Stella Čapek states that the tenets of these different rights movements translate into the environmental justice movement through several key claims about people's right to the following:

> (1) accurate information about the situation; (2) a prompt, respectful, and unbiased hearing when contamination claims are made; (3) democratic participation in deciding the future of the contaminated community; and (4) compensation from parties who have inflicted injuries on the victims.
>
> (1993, p. 8)

She also points out that those within the environmental justice movement work to seek solidarity with others impacted by justice issues. Dorceta Taylor calls this a master frame, and says that the "environmental justice frame has emerged as a master frame used to mobilize activists who want to link racism, injustice, and environmentalism in one frame" (2000, p. 514).

Regardless of the movements from which it was born, environmental justice has its origins in considerations of people in relation to environmental concerns, specifically the relationship between race and poverty and the distribution of waste and industry (Bryant & Mohai, 1992; Sze & London, 2008; Walker, 2012). As such, definitions and framing within the field focus on two main questions: (1) what is the condition of inequality?; and (2) what is considered just and/or unjust? Further, these considerations are frequently framed in relation to an objective and the case at hand itself, making environmental justice a movement and field that are multidimensional in nature (Dobson, 1998; Walker, 2012). In a later section in this chapter, I will detail some of the critiques of the original framing of the field and describe scholars' call for an expansion of these fundamental considerations.

In the United States, environmental justice has several widely agreed-upon characteristics (Walker, 2012), including considerations of politics of race, justice to people, environmental boundaries, distributive justice, blame and responsibility, scales of analysis, and the government and its agencies. Throughout this book, I explore these different dimensions and their relation to labor along the scales of the agricultural system in the United States—from small-scale organic agriculture to large-scale conventional agriculture. In their implications for work with farmworker justice specifically it stands to reason that the orientations most closely related to injustice and cultural oppression and those that recognize the political oppression of certain populations are most relevant to consider and explore.

Categories, organizations, and the place of labor

While environmental justice as a field grew largely out of the antitoxic and civil rights movements, there are some interesting distinctions in the types of organizations and organizing between these groups. Within the environmental justice movement, the organizations are characterized by a higher overall amount of coalition activity, which also turns them more easily toward cooperative relations with labor organizations (Agyeman, 2005; Obach, 2004). Brian Obach (2004), in *Labor and the Environmental Movement: The Quest for Common Ground*, points out that by highlighting the previously under-noticed social impact of environmental hazards and those who work in direct contact with them in the early 1990s, environmental justice groups paved the way for increased collaboration with labor organizations. He goes on to discuss a continuum upon which environmental organizations exist, at one end, those organizations focused on land use (protection of wilderness and species) and, on the other, those focused on the effects of environmental hazards on human health and well-being. According to Obach, those environmental organizations that are most successful are those that rest somewhere in the middle of this continuum, focused equally on environmental and social impact concerns. Julian Agyeman (2005) employs a similar approach as he categorizes organizations based upon their commitment to both the principles of justice and those of sustainability. Both authors make central their claims that a balance between environmental- and justice-based concerns is ideal.

Obach (2004) creates another useful continuum that contextualizes the work of environmental organizations and their relation to labor, particularly in the United States. At one end of this continuum is a sole focus on environmental issues, and at the other is a singular focus on social and workplace issues. More specifically, he places land use organizations and environmental organizations at the environmental end of this continuum, with environmental organizations extending further to the middle. At the other, he places business unions and social unions, with social unions extending further to the middle. Also spanning the middle are environmental justice and toxics organizations, which focus in a more balanced way on solving both environmental and social/workplace issues. For this reason, this last grouping of organizations has pushed those at the margins of the continuum to expand their range, and to take new dimensions into consideration.

Philosophical and theoretical foundations of environmental justice

To give further context to the discussion of environmental justice, it is important to get a sense of the landscape of the foundational literature that frames justice-related thinking more broadly. Table 1.1 explains briefly some definitions of theories of justice commonly discussed by environmental justice scholars and also includes a survey of justice-related literature to offer alternatives to distributional justice and utilitarianism, the common focus areas of the field of environmental justice.

Peter Wenz (1988) argues that environmental justice currently focuses on two main theories of justice: distributive and utilitarianism. Wenz states:

> According to utilitarianism, those actions and policies are justified which produce the greatest good. It is objected that such actions and policies will likely fail to do justice to individuals, or to groups of individuals. Many will fail to get what they are due, and this is unjust. It has been urged against utilitarianism, for example, that happiness may be maximized by making relatively few people very happy, at the expense of everyone else. (ibid., p. 181)

In addition, Wenz explains that utilitarianism at times requires that some people in society be treated unjustly, particularly when equality is not possible. It is when that requirement disadvantages the marginalized or subaltern in a society that a fundamental conflict occurs, particularly with an espoused tenet of the environmental justice movement: to acknowledge and give voice to those disenfranchised by the systems within which we exist. If the movement relies too heavily on these two theories of justice without considering the applicability of other theories such as productive, restorative, transformational, and transitional

Table 1.1 Definitional review of relevant theories of justice

Theory of justice	*Definition*
Distributional	The two principles of justice envisioned within a distributional framework are: (1) personal liberty that is compatible with the liberties of others; and (2) wealth that is beneficial to the least advantaged and attached to those positions within society for fair equality of opportunity (Barry, 1989; Rawls, 1971; Sen, 2009).
Productive	A form of justice based on addressing the political-economic structures that produce environmental problems, aimed at creating a system within which we focus on causes rather than symptoms. Emphasizes participation in the decisions through which environmental burdens are produced. Characterized by a movement from a "not in my backyard" to a "not in anyone's backyard" political frame (Faber, 2008).
Utilitarianism	There are two forms of utilitarianism: (1) act utilitarianism, which is the theory that a morally right action is one that in the existing situation will produce the highest expected social utility (commonly rephrased as "what does the most good for the most people"); and (2) rule utilitarianism in which the morally right action is simply an action conforming to the correct moral rule applicable to the existing situation, defined as the rule with the highest expected social utility (Harsanyi, 1985).
Restorative	Restorative justice begins with three questions: (1) who has been hurt?; (2) what are their needs?; and (3) whose obligations are these? In a quantification or assessment of loss or harms, this model focuses on the needs of the victims and the responsibility of the offender versus retributive justice, which follows the abstract principles of the law. For restorative justice to work, there needs to be an easily identifiable victim (Johnstone, 2007).
Transformative	Transformative justice focuses not only on the victims and the offender, but also on the broader community. This is a healing form of justice rooted in the Quaker philosophy of forgiveness and working backward from a future desired state. Transformative justice is a systems approach to justice, allowing one to see an offense as both relational and educational for both victims and offenders (Morris, 2000).
Transitional	Transitional justice focuses on communities transitioning after conflict. Key elements to transitional justice: identifying individual people's level of responsibility, preserving and enhancing peace, halting human rights abuses, investigating past crimes and proposing sanctions for those crimes, and fostering reconciliation, to name a few (Roht-Arriaza & Mariezcurrena, 2006).

justice, it misses an opportunity to truly rethink the way that we live and interact with one another, as well as the natural world. Wenz offers an alternative model he calls the concentric circle perspective, which draws from other theories of justice and explores notions of obligation, justification, and positive and negative rights, which will be explored further in the next section.

Based on notions of distributional and utilitarian justice, the environmental justice master frame examines the balance between people's capabilities and their access to things such as life, freedom, resources, happiness, well-being, and liberty—meaning people's ability to take full agency of their own lives and holistic health. At its base, distributive justice is about equity in distribution of social goods. In *The Idea of Justice*, Amartya Sen (2009) addresses the complexity of understanding that all people actually are created equal with the stark realities of the practical implications of that very statement—insomuch as if that is true, then all people should have an equal chance (a fair chance) to access those things and live the lives they want. Complicated by the imposition of social stratification and the history of humans as creatures who have tended toward separation rather than integration and competition rather than collaboration, this same philosophical conundrum is central to the field of environmental justice. Considerations of justice are complex, and the conditions for achieving justice many times are at odds with the very ways that society is structured.

John Rawls (1971) was a key figure in shaping ideas about distributional justice in his book, *A Theory of Justice*. He discussed the idea of a veil of ignorance, wherein a person has no idea what role he/she plays in society, nor what status he/she has in the said society and must conceive of a social contract that contains principles and values for the good of all. From Rawls' perspective, this would allow for a system that is "fair" even to those who are traditionally the most disadvantaged in a society. From Rawls' viewpoint, these basic liberties are: freedom of thought, liberty of moral/religious/philosophical conscience, political liberty, freedom of association, freedom necessary for liberty and integrity (freedom from slavery with an ability to choose one's own occupation), and the rights and liberties covered by law. The two principles of justice he envisions are: (1) personal liberty that is compatible with the liberties of others; and (2) wealth that is beneficial to the least advantaged and attached to positions of fair equality of opportunity.

In contrast, while Michael Walzer (1983) acknowledges that this is a useful equation for considering what people might do absent other knowledge, he states that it is not particularly useful in real life. He states that, among other reasons, this is because social meanings change the values of things over time and thus individuals create their own distributive sphere through which they decide what is right, fair, and just. Further, when considering what a notion like everyone getting their 'fair share' would look like, he asks the question, "Fair shares of what?" (ibid., p. 79), pointing out how difficult not only deciding on what is fair would look like, but also the complexity of identifying the social goods to which that concept would even apply.

Martha Nussbaum considers the key questions, "What are people actually able to do and to be? What real opportunities exist to them?" (2011, p. x), which dovetail with the examination of labor as a dimension of the environmental justice movement. She argues that each person should be an end in themselves, rather than focusing on averages as seen in utilitarianism, and that justice for said individuals should be focused on the expansion of their individual choice and freedom. In doing so, she frames a capabilities approach to justice, centered on three areas: (1) internal capabilities (those traits and abilities that we can develop); (2) basic capabilities (those that are innate to an individual); and (3) combined capabilities (those that are developed, or not, because of an individual's specific political, economic, or social situation). In doing justice work, the area of combined capabilities offers the most opportunity for systems change and influence. Among the capabilities to consider, Sen and Nussbaum build upon one another's work, and identify several central capabilities: life; bodily health; bodily integrity; senses, imagination, and thought; emotions; practical reason; affiliation; other species; play; and control over one's political and material environment. To be fully functioning, one must both have the freedom to choose as well as the potential to actively realize these capabilities.

While the historical dimensions, scale, and theories of justice employed within the environmental justice movement deserve consideration, other areas do as well. The question: "Justice to what end?" must be carefully considered. For example, common practice within the field argues that we must document the number of injustices and problems facing marginalized communities, and in doing so highlight the structural and systemic foundations of injustice. All too often in practice, doing so results in localized, short-term gains, rather than changing the system and structures themselves.

Building on these broader ideas about justice, Gordon Walker (2012) gives a useful overview of the different facets of environmental justice. The first, justice, is normative in nature and focused on how things ought to be, giving a standard by which we can determine whether something is right or wrong. The second is evidence, which is descriptive in nature and centers on discovering how things are and/or examining the facts associated with them. Third is process, which considers why things are the way they are, and is explanatory in nature. He uses this framework for claims around environmental justice, and further urges that research and literature in this area are most productively focused when looking at evidence, or descriptions of conditions of difference or unevenness between different groups of people.

The environmental justice summit: defining a movement

While the environmental justice movement began in its infancy in the 1980s with a growing awareness of the impact of toxics on communities of color, it was the early 1990s that saw a true movement take shape. With the 1990 University of Michigan Symposium on Race and the Incidence of Environmental Hazards, which gathered activists and academics to discuss growing momentum and discourse around these topics, and the 1991 First National People of Color Environmental Leadership Summit, which gathered people from all across the United States to discuss the meaning of both environment and justice, the new movement began to take shape. Throughout these conversations, the environment was framed as the place where people live, work, and play, and it became clear that there was radical disparity between access to a healthy environment across different classes and races of the U.S. population (Bryant & Mohai, 1992; Foreman, 2011).

Representatives at the summit varied from those involved with civil rights, antitoxics, academia, the labor movement, farmworker advocates, traditional environmentalists, Latino, Asian-Pacific, and Native American struggles, all of which are present in the ensuing principles (Cole & Foster, 2001). At the summit, held October 24–27, 1991, these delegates adopted 17 Principles of Environmental Justice. Part of the introductory paragraph to these principles states as their purpose, "to ensure environmental justice; to promote economic alternatives which would contribute to the development of environmentally safe livelihoods; and, to secure our political, economic and cultural liberation that has been denied for over 500 years of colonization and oppression." Among the 17 principles are ideals that posit that public policy should be based on its benefit to all people and thus free of discrimination, an affirmation of the right of all people to political, economic, cultural, and environmental self-determination, the right of equal participation in all stages of decision-making, the right of all people to a safe working environment that does not force a decision between work and one's safe livelihood, an opposition to the destructive practices of multinational corporations, and an education about social and environmental concerns that is culturally competent and specific (Delegates to the First National People of Color Environmental Leadership Summit, 1991).

The second summit, held in October 2002, celebrated triumphs that had occurred within the movement since the first summit, and explored obstacles to movement growth and efficacy. One source states that:

> Prioritization of issues became a topic of debate among participants along with how to best utilize funding sources and grant monies. How can bureaucratic agencies of the federal government, academic institutions and grass roots activists work together to address our ultimate goals and objectives in the EJ movement?
>
> (SRIC: Voices from the Earth – 2003, 2003)

In addition, concerns emerged on behalf of the farmworker justice community that the efforts of the movement were not moving the needle on issues related to farmworkers, and that "Farm work is the second most dangerous occupation in the United States. Farm workers suffer from the highest rate of chemical injuries of any workers in the United States" (*The Black Commentator*, n.d.).

The principles established by the summit, combined with the philosophical foundations of justice and their interplay with the environmental justice movement, form a framework for viewing the plight of laborers within U.S. agriculture, no matter the scale, as an environmental justice issue. While contextual process claims arising from specific situations may differ across the continuum of agriculture from large-scale conventional to small-scale organic, the structural process claims, or systemic inequality throughout our agricultural system as illustrated by history and policy, are pressing on the laborers all along this scale and creating not only unjust working environments, but a system that is unequipped to heal itself. That, in combination with what I discuss in the next section regarding broader calls in the environmental justice movement for recognition, capabilities, and participation, is grounds for further exploring the exploitation of agricultural labor as an environmental justice issue and specifically what and how laborers themselves can and must be involved in changing this context.

Structural and systemic perspectives on environmental justice

Like many fields that challenge dominant paradigms, environmental justice has encountered critics claiming that it focuses too much on discrimination of marginalized populations, who perhaps are not targeted at all, but rather are drawn to ecologically damaged and polluted areas (Wenz, 1988). This criticism of the field maintains that too much emphasis is placed on the wrongdoing imposed upon minority and marginalized populations. However, this critique fails to recognize the larger political-social-economic context. Julie Sze and Jonathan London point out in their article, "Environmental Justice at the Crossroads," the need for expansion in the field in two areas: the first being new populations and problems, and the second new places and sites of analysis (Sze & London, 2008). They state:

> In its earliest incarnations, environmental justice research was concerned primarily with the toxic and hazardous waste impacts in low income and communities of color. It has since expanded the kinds of environmental inequalities being studied in areas such as transportation, health, housing, and smart growth/land use, water, energy development, brownfields, and militarization. Recent research also has both broadened the scope to include other racial and ethnic groups … and disparities associated with gender and age.
>
> (ibid., p. 1337)

John Callewaert (2002) suggests that the environmental justice field can address critiques through inclusion of the historical perspective and broader socio-economic considerations. He states, "Another aspect of the multi-dimensional process is the significance of historical precedence in developing situations of environmental injustice. The importance of highlighting and articulating such issues is critical to the acceptance and resolution of environmental injustice claims" (ibid., p. 265). He also points out that a shortcoming of traditional environmental justice research is its proclivity to focus on a singular event, policy, or health effect and its impact on a marginalized population. He suggests that this dialogue should be broadened by further exploration of the historical dimensions of a community or situation. Callewaert asserts the following through his case study findings:

> In all three cases of this study the communities existed prior to the development of disputed industrial facilities. What is more significant is that each of the communities developed as communities of color because of earlier housing discrimination practices. Such discriminatory practices created a climate allowing undesirable and in some cases inappropriate industrial neighbors, that would have never developed in wealthier white communities. (ibid., p. 263)

Callewaert expands the scope of the field by suggesting that we include a broader consideration of how communities develop, and thus illuminate how inherently racist and discriminatory practices commonly prevail.

In addition to the expansion of the scale of time used to analyze environmental justice issues, scholars in the field also challenge the geographic scale at which traditional environmental justice research is done, which has been predominantly local. While useful, a consideration of issues at the neighborhood and small community level only allows the field to progress to a certain point. An expansion in geographic scale would not limit research from happening at the local level, but would connect these local efforts with broader, systemic issues that could begin to be addressed at the policy and institutional levels.

Logically, if the scale of environmental justice were to expand, it leaves one pondering what theory (or theories) of justice are most helpful to consider. When looking back at Table 1.1, the theories of productive, restorative, transitional, and transformational justice may have interesting applications within the environmental justice movement. Notably, simply using distributional justice as a guide becomes problematic when considering how to further the field. While not specifically looking at environmental justice, Iris Young (1990) points out that justice needs to focus more generally on the elimination of institutionalized domination and oppression of those who represent difference. In another piece, Young and Christian Hunold point out that utilitarian and distributive notions of justice are insufficient, building upon her earlier assertion and also emphasizing the important of participation (Hunold & Young, 1998). They also follow that critics of democratic participation-based models of change often express a concern that in these systems people resort to simply advocating for their own interests. These considerations have important implications for the field of environmental justice, as well as deciding which theories of justice are most useful in considering how to make systemic change.

Pellow and Brulle suggest we must "take seriously the obligation to propose new directions for society to heal itself and produce more just and sustainable forms of production" (2005, p. 296). While the movement has had both a reporting and documentation mindset and one truly rooted in social change, many critics and scholars highlight the necessity of clarity of focus around the latter. Many ways exist to involve people in actively creating change to positively affect their own and others' lives, as illustrated through many of the successes within the environmental justice movement. Further, considerations of the scale at which we examine environmental justice issues and engage communities in social change are also in question. Related to this, Giovanna Di Chiro discusses this using the term "living environmentalism," positing:

> Innovative environmental justice and reproductive justice coalitions articulate people's concerns about their families' and communities' access to social reproduction – the maintenance and sustainability of everyday life and earthly survival made all the more difficult by global economic and environmental crises. These coalitions are generating dynamic, living environmentalisms that may well compel people to join together and take stronger action to curb problems as big as global warming.
>
> (2008, p. 294)

These scholars are touching on a major idea—one that moves the EJ movement from one that is situated in making locally based, distributional change, to one that is able to scale for larger, systemic change that involves the very people the change impacts.

Recognizing the limitations of traditional concepts of environmental justice, David Schlosberg's work offers a paradigm for addressing these issues, claims, and injustices. Schlosberg (2007) notes that both self-described environmental justice movements and theories of environmental and ecological justice use the word "justice," but sometimes in very different ways. He discusses theorists who not only identify distribution as the means toward justice and equality, but also those who examine the causes of the present misdistribution of goods, services, and harms. He explores the multifaceted nature of what people consider to be "environmental justice," ranging from specific toxics issues to deeper issues of race and inequality. This discussion sheds light on why people are often confused about the concept of environmental justice and find it somewhat nebulous.

Schlosberg (2004; 2007) argues that a unified vision and definition of environmental justice expanding beyond notions of distributional justice are needed so that the work can occur on a global scale. He points out that a theory and practice of environmental justice necessarily include distributive conceptions of justice, but must also embrace notions of justice based on recognition, capabilities, and participation (Schlosberg, 2007). Expanding environmental justice to the global scale, Schlosberg states:

> Certainly, at the center of the recent protests against global financial and trade institutions, against the globalization of the food system, and for indigenous rights, is the issue of equity; economic or distributive injustice

> is a key and constant rallying cry. The most basic critique is that the currently favored model of development increases and exacerbates inequity, both between the North and the South and between elites and the impoverished in southern nations – again, not just in economic goods, but in environmental goods and bads as well.
>
> (2004, p. 523)

He notes as well a lack of connection between the environmental justice movement and other movements is centered in related but divergent notions of justice: economic justice, political movements, social justice and change, and others. Furthermore, Schlosberg exposes that the range of justice traditionally considered does not extend beyond human life, that "nonhuman nature is simply beyond the bounds of the relationships that can be based on justice" (2007, p. 104). His goal is the development of a broad yet integrated notion of justice that applies to both relations regarding environmental risks in human populations and relations between human communities and non-human nature. When scaling out notions of environmental justice, to not consider these recognition-based and participatory elements is the downfall of the environmental justice movement in Schlosberg's opinion.

Schlosberg states the following, defining recognition and participation:

> The argument is that the justice demanded by global environmental justice is really threefold: equity in the distribution of environmental risk, *recognition* of the diversity of the participants and experiences in affected communities, and *participation* in the political processes which create and manage environmental policy.
>
> (2004, p. 517, emphasis added)

By recognizing group difference and allowing all groups the rights of participation, one seeks a truly 'just' form of justice. Schlosberg also argues for the need for community investment, stating, "the demand for this type of authentic, community-based participation comes out of experience of disenfranchisement – the combination comes out of misrecognition and political exclusion" (2007, p. 68). Environmental justice from this vantage point and authentic stakeholder participation, when applied to rights for farmworkers, holds untold possibility for advancing the cause of justice for those who toil in our fields to provide us with nourishment. It would not only move their concerns to the forefront, but their voices as well, by involving them in a central way in the fight for justice. Axel Honneth explains that participation has expanded and changed over time as our notions of who should participate and how have evolved. He states:

> In the beginning, only those who could demonstrate a certain measure of income or property had a positive claim to participation in the process of political will-formation. These previously status-bound rights to participation only became a distinct class of universal human rights once these

> rights were partially expanded and carved out and the legal-political climate had changed to such a degree that there were no longer any convincing arguments with which to oppose the demands of excluded groups for equality.
>
> (1995, p. 116)

Recognition necessitates that one explores how communities are unfairly affected by different environmental harms. In doing so, one acknowledges, sees, and hears communities and their people who many times have been ignored both by the larger environmental movement, and their own supposedly representative governments. An expansion to considerations of participation center on community involvement in the design and ongoing oversight of environmental risks that impact them (Schlosberg, 2003). Toward this end, he states, "It is not just that political and cultural institutions create conditions that hamper equity and recognition, but that both distributive inequity and misrecognition hamper real participation in political and cultural institutions" (ibid., p. 86). It is clear that there are opportunities to expand not only the scale of environmental justice, but also how it is enacted in practice—rather than a community of scholars doing work *for* communities, a group of collaborative relationships that *include* community involvement and naming of historical wrongs and challenges and the root causes behind them.

Walker builds upon Schlosberg's work, noting that "qualitative methods have also been used to reveal the values and subjectivities that underpin how different environmental goods and bads are understood by different social groups" (2012, p. 62). He further distinguishes different claims within the environmental justice movement, separating them into two categories: contextual process claims and structural process claims. Contextual process claims are rooted in a specific situation whereas structural process claims are centered on how society works and how "uneven environmental outcomes are the consequence of the systemic structuring of social relations" (ibid., p. 64). He reinforces Schlosberg's call to include distribution, recognition, and participation, stating that although they are each distinct justice concepts, the overlap and interrelation between them are what constitutes full justice; and discovering where a process unearths inequalities contributes to one's sense of the whole and the various possible points of intervention.

It is important to note, as Daniel Faber and Deborah McCarthy point out, "Although most organizations or *movements* for environmental justice are distinct from one another in a number of rather profound ways, it should be emphasized that *all are united in the larger struggle for ecological democracy*" (Faber & McCarthy, 2003, p. 46). They discuss how each of the organizations seeking to connect the problems faced by historically marginalized communities into the tapestry of larger democratic participation and change face the challenge of examining not only the current political context of their issues, but the historical reasons for the subjugation. Further, they note that there are a number of networks of otherwise separately operating organizations evolving over time as the environmental justice movement grows for the purposes of centralizing and leveraging their strength.

This ties into political economic considerations within environmental justice, focusing on the systemic structures within which injustice exists. In its environmental racism roots, one must consider the economic, political, legal, and social forces that impact environmental decision-making processes and outcomes (Cole & Foster, 2001). If the focus of the movement is solely localized distributional justice, then the possibility of broader systemic change is not realized. Luke Cole and Sheila Foster explain, "A view from the ground allows us to see the many dimensions of power struggles, the relationships of actors within these struggles, and the role of the legal and regulatory framework in structuring those relationships" (ibid., p. 12). Therefore, it is not enough to simply address unequal distribution, but we must also consider the processes by which that inequality came to be (Heynen et al., 2006; Walker, 2012; Watts & Peet, 2004).

John Green, Eleanor Green, and Anna Kleiner note of this as well, pointing out that organizations and institutions such as the market, the state, and civil society are at the center of whether people can access the very pathways for their own improvement (Green et al., 2011). Considerations of justice, with this in mind, must focus on the systems that create injustice, as well as remedying the outcomes of those processes. Ryan Holifield, Michael Porter, and Gordon Walker also discuss the ways that systems and structures limit the ways that environmental injustices can be and are expressed and addressed (Holifield et al., 2010). They cast the state in relation to its role in the racialization of these processes, pointing out the focus on race versus class, lay versus expert knowledge, and particular versus abstract knowledge in the pursuit of change. Further, they state that environmental justice "needs to address a wider range of inequalities than it currently does" (ibid., p. 71). To break down these dichotomies and to be able to expose the root of the issues requires involvement of the people they affect, and to involve them necessitate the development of strategies for organizing.

When considering strategies for organizing, Rose (2000, p. 116) states, "[it] requires the right marriage of strategy and circumstance" and illustrates through case studies how coalitions rely upon a recognition of external circumstances, opportunities, and limitations and how to leverage them appropriately in the context of the issue at hand. Strategies become more complex as more players are added, and when many organizations or stakeholders are at the proverbial table, it can be complicated to navigate through the various beliefs, values, and desires represented. However, when it comes to complex issues such as those facing farmworkers, it is increasingly apparent that this marriage between circumstance and strategy is imperative. This leaves one asking the question, how? How does one create relationships necessary for justice and social change, when others enter the conversation with distinct, diverse interests at the forefront of thinking? Expanding the scale of focus in addition to a systemic analysis of the causes of injustice, including the historical context, makes coalition building not only desirable but necessary to understand complex issues and create strategies for social change.

Key considerations of environmental justice

Environmental justice is a movement rich in examples of organizing around issues to make change in policies, procedures, and organizations that create unequal conditions by which the subaltern disproportionately bears the burden of ecological hazard and/or is actively excluded from the processes that shape their own livelihood. However, while scholars and activists within the field have called for change and expansion, one is left considering what next steps are necessary to advance the field. Walker (2012) suggests that we come to environmental justice concerns from different claim points. In doing so, we can focus on a set of questions: (1) who is responsible?; (2) why is it unfair?; (3) what would be fair?; and (4) who is impacted and why? The answers to these questions, in addition to the factual information gathered in environmental justice claims, leaves us to more appropriately address and consider approaches to alleviating, or at least confronting, injustice.

If we are to create sustainable change, I agree with the scholars above who focus on participation and true integration of those capabilities and strengths of communities. Bringing more voices to the table also prompts consideration of how fairness and justice matter across difference and impact how we frame problems and solutions that are representative of a diversity of perspectives, and are aimed at making sure we are accounting for those least advantaged. The fields of coalition building and asset-based community development offer models for how to do this, and enrich the narrative asserted by the environmental justice scholars above that align with expansion of the scope and considerations of the field.

Drawing from coalition building and asset-based community development

If participation, capabilities, and recognition are to be realized within environmental justice, it is useful to use the frameworks provided by the fields of coalition building and asset-based community development. According to many scholars and practitioners, the first step in building solidarity and coalition is authentic, innovative, democratic leadership. Richard Couto (2002) lists values, inclusiveness, initiative, and creativity among the most important elements of such leadership—necessary ingredients to making cross-interest relationships work effectively. Some key steps include agreeing upon a shared set of values, articulating strategies for inclusiveness, depending upon members for initiative when a need speaks to their strengths or resources, and leveraging the creativity in groups rather than shying away from it. In doing these things, diverse groups of organizations and individuals can begin to find a common foundation upon which to build strategies for the problem they hope to address together.

Of course, making sure that the appropriate voices are included in the dialogue and that the context for the circumstance is fully explored are important considerations as well. Couto (1999) reminds us that when forming relationships and considering strategies, human worth must exceed market and labor values. He states:

> The roles of community-based mediating structures extend beyond nursing the wounds that market capitalism inflicts. They include efforts to revive and sustain the democratic prospect of increased amounts and improved forms of social capital. These efforts express the value and worth of groups and individuals that are redundant or marginal to the labor needs of the market.
>
> (ibid., p. 72)

Rhonda Phillips and Robert Pittman discuss social capital and capacity as central to the community development and coalition building process, in their book, *An Introduction to Community Development*, stating, "If citizens cannot plan and work together effectively and inclusively, then substantial community progress will be limited" (2009, p. 49). Social capital concerns the connections between and among people, and our capacity to use relationships as a resource to accomplish and work toward our goals (ibid.). The implications of these realizations for farmworkers are immense, given the political economic terrain (discussed thoroughly in Chapter 3), emphasizing a need to create avenues to build social capital and capacity for strategy enactment.

A thorough consideration of the types of public participation is needed when figuring out how to encourage and support authentic involvement from multiple stakeholders. Sherry Arnstein (1969) proposed a classic framework for considerations of participation called "Eight Rungs on a Ladder of Citizen Participation," which ranges from forms of nonparticipation (manipulation and therapy) to those of tokenism (informing, consultation, and placation) to those of citizen power (partnership, delegated power, and citizen control). Gary Green and Anna Haines (2001) build upon this work to identify four types of public participation: public action, public involvement, electoral participation, and obligatory participation (Table 1.2). Giving stakeholders different options for entering the process and participating as they choose is crucial in reaching multiple needs and addressing as many interests as possible.

Further, as Ife and Tesoriero (2002) point out, expanding notions of participation to include a wider range of perspectives necessitates a period of trust-building between the groups working together, in addition to a clear process for vetting concerns when that trust is broken. Participation is messier when diversity is valued rather than silenced, expanding the parameters of concern and areas for action. The venues for participation must address this and those involved have to be open to the process when there are no easy answers, building upon their trust that everyone has another's best interests at heart. This is reminiscent of Rawls' (1971) veil of ignorance, wherein a person has no idea

Table 1.2 Definitions of variations of public participation

Public participation type	*Definition*
Public action	When the public (whether through an individual, group, or organization) exerts influence on the government to modify existing policies and practices.
Public involvement	"Activities initiated and controlled by government for administrative purposes" (Green & Haines, 2001, p. 36)
Electoral participation	Voting in elections (viewed as a minimalist form of participation) is an indirect way of communicating one's preferences about policies and debates
Obligatory participation	Obligations of citizenship, including paying taxes, serving on juries, etc.

Source: (Green & Haines, 2001).

what role he/she plays in society, nor what status he/she has in said society and must conceive of a social contract that contains principles and values for the good of all. It is then expanded by Schlosberg's (2007) call for the inclusion of participation in conceptions of justice.

According to Couto (1999), navigating this diverse field of interests requires community-based mediating structures and their interplay with democratic theory and social capital, although it is unclear at what scale these community-based structures begin to fall apart. Does this only work in small groups? Local communities? Large, complex businesses and organizations? Or, can it be realized at a regional, national, or even international scale?

In addition to answering these questions, a framework is needed for understanding and encouraging community change. While many models of community development exist, critical theory, considerations of power and privilege, and broader theories of justice resonate most closely with asset-based community development. In *Asset Building and Community Development*, Green and Haines (2001) offer asset-based community development as an alternative to the more traditional needs-based community development, which focuses on the detrimental aspects of a community and the development of strategies for changing them. Traditional environmental justice practice resonates from this needs-based model, which Schlosberg (2007) points out as a hindrance to the expansion and sustainability of the movement. Asset-based community development is:

> The reverse of the conventional approach. The idea is to build capacity within a community – to build and strengthen a community's assets. In contrast to focusing on problems and needs, this alternative approach … is focused on a community's capacity rather than on its deficits.
>
> (Phillips & Pittman, 2009, pp. 39–40)

In alignment with Schlosberg's (2004; 2007) discussion of capabilities and recognition, asset-based community development allows the full perspective of what a community and its residents *can* do with their strengths combined. Rather than mapping needs like poisoning, dropouts, broken families, and crime, we begin to map organizations and individuals like schools, parks, churches, and cultural groups (Green & Haines, 2001). Doing this not only builds morale, but also allows the community to start from a place of agency and empowerment in solving their own concerns for sustainable change. After inventorying assets, a community can more readily begin to form coalitions, collaborative relationships, and leverage their social capital to accomplish their goals.

Finally, all of this work at its best should focus on the process in addition to the outcome. As stated by Jim Ife and Frank Tesoriero (2002), community development is about more than bringing people together and solving a problem. "Group process, inclusiveness, building trust and developing a common sense of purpose are all critically important in community building" (ibid., p. 185). This shift of focus allows people engaged in the process to truly participate and consider the roles and viewpoints of others. For this to happen, having structures in place through which people can participate and voice their concerns is key. In considering farmworkers and the other stakeholder groups invested in their just treatment, allowing multiple forms of participation and having a keen eye toward the process, including building trust, are important ways, among others, to build coalition and ensure success.

Dimensions of environmental justice

Scholars, practitioners, and activists alike would agree that there are multiple dimensions of consideration in environmental justice claim-making. There are the social dimensions, which underscore the interplay of human dynamics in environmental justice, as well as the environmental dimensions that serve as the basis for consideration in equality. Many critics of the movement cite the pre-eminence of human approaches to change as a shortcoming of the field, particularly in addressing the systemic dimensions of environmental and ecological hazards themselves (Benford, 2005). On the other hand, proponents of environmental justice reason the necessity of this dynamic, particularly because of the power dynamics inherent in the environmental movement itself, because "social justice and environmental sustainability are not always compatible objectives" (Dobson, 2003, p. 83). When Robert Benford (2005) reviewed environmental justice websites, he discovered that there were over 50 issues that the movement claims to be working toward ameliorating and/or including, from deforestation to human rights and from disability rights to sovereignty to poverty. Ranging from small-scale to enormous, systemic issues, the movement is currently lacking a central focus.

Walker (2012) describes the social and environmental dimensions of the environmental justice movement in a way that makes them simpler to understand. Among the social dimensions, he lists expected dimensions such as race, ethnicity, class as well as some that are not part of classic environmental justice considerations as rooted in environmental racism such as single parent families, gender, indigenous peoples, and future generations. The environmental dimensions of the movement vary considerably from traditionally considered notions of toxics, landfills, air and water pollution, and incinerators to broader considerations of access to healthy food and green space, agriculture, trade agreements, and land reform. The interplay of social and environmental dimensions make for a rich, yet vast and ever-changing landscape for environmental justice as a movement and a field of scholarly research.

Holifield, Porter, and Walker (2009) also point out four dimensions of environmental justice which they believe are in need of further and deeper engagement. These are: (1) spatiality (environmental justice as a field focused on distributive outcomes mapped in geographical space ignores the other dimensions of justice such as participation, recognition, and capabilities); (2) the role of the nonhuman in environmental inequality (our interaction as a field with these actors such as air, water, land, chemicals, waste, equipment, etc. determines and influences how we define and measure justice, risk, and inequality); (3) gender (both differences in how men and women experience environmental inequality and vulnerability, as well as the gendering of governmental and non-governmental entities, which are predominantly influenced by men due to the large percentage of positions occupied by them); and (4) the state (how policies, programs, and laws affect environmental practices and address environmental inequalities and the state as a racializing actor and its role in the creation of race as a central factor in modern life, regardless of its intent to do so).

Labor in agriculture as an environmental justice issue

Environmental justice concerns are both situated in time and place, making them contextual in nature, complicating the possibility of a universal definition (Stephens et al., 2001). However, there are useful frames from which we can begin to understand, document, and examine environmental injustice, creating a platform from which to solve problems. The concepts of justice as distributive, procedural, and recognition-based together create a framework from which to view laborers across all scales of agriculture in the United States. From this framework, conventional farmworkers encounter environmental risks aplenty in the course of their daily life, especially with regard to pesticides. In both conventional and organic agriculture, justice issues are linked to larger questions about the right of people to work in a safe and healthy environment without sacrificing the basic needs of a vibrant livelihood. These include making a living wage, access to affordable housing, access to education, gender equality, the ability to organize, and freedom to work where one chooses and to report violations without fear of retribution. Within the larger political economic system of the United States, organic agriculture (particularly small- and family-scale) is marginalized due to lack of the possibility of full participation in the market system and the inaccessibility of benefits provided by the government to agricultural entities, more fully explored in Chapters 3 and 5. The interplay of these policies and their creation of both localized and systemic structures through which farmworkers are marginalized will be covered in later chapters. Here, I introduce the main issues facing laborers across different scales of conventional and organic agriculture in the United States, which are detailed in Chapter 2.

Conventional agriculture

Injustice faced by laborers within agriculture has been well documented over time, with more specific attention paid to it in the last two decades. While there are a number of issues impacting farmworkers within conventional agriculture, pesticide exposure and pesticide-related illness are the most well documented. As members of society become more attuned to what they put into their own bodies, researchers have studied the impact that concentrated exposure to these chemicals has on farmworkers. In *Fields of Poison* (Reeves et al., 2002), the Pesticide Action Network North America (PANNA) reports on the use of pesticides and regulation lapses in the state of California, and makes recommendations to the Department of Pesticide Regulation. By highlighting the types of pesticides most frequently used and causing harm, the crops that use these poisons, and data regarding regulation oversight, PANNA offered suggestions focusing on an "urgent need for safer agriculture and better worker protections" (ibid., p. 7). Chief among these nine recommendations are protestations about the importance of reporting, regulation, and enforcement, making more central the understanding that what consumers fear will harm them when consuming food is exactly what laborers in our fields face each day as they go to harvest.

On the tail of recommendations like those put forward by PANNA, the Center for Disease Control and Prevention's (CDC) National Institute for Occupational Safety and Health published *Pesticide-Related Illness and Surveillance: A How-To Guide for State-Based Programs* (CDC–NIOSH, 2005). While the CDC had been providing financial and technical support for the surveillance of pesticide use and illness since 1987, this report signaled a turning point for the provision of free and easily accessible resources for agencies interested in monitoring practices and assisting farmworkers, also compiling the most harmful and numerous injustices against them. The comprehensive guide includes: (1) information about the importance of pesticide poisoning surveillance; (2) mechanisms to improve reporting of cases to surveillance programs; (3) methods to investigate reported cases; (4) guidance on using the case definition; and (5) additional resources on pesticide toxicology, pesticide usage, governmental partners, and surveillance (ibid., p. iii).

Another report that came out in late 2011 is *Farmworkers in the Southeast* (Sologaistoa, 2011). It builds upon the structure and methodology advanced in the PANNA report, but focuses on the regional rather than the state level. It illuminates how difficult it can be to gather information about the farmworker population due to its migratory patterns, but also shows that it is possible to work together across state boundaries to talk about the issues facing farmworkers on a larger scale. It shows that by working together, we can sometimes gather more representative information about this mobile and shifting population. In this report, Erin Sologaistoa structures the recommendations for various audiences, from government regulatory agencies to farmworker organizations. This report demonstrates the value of scaling up data collection and subsequent calls to action, suggesting that research coalitions are a valuable possibility to explore for the future.

Finally, migrant and immigrant labor is critical to consider. According to some estimates as many as 80 percent of the agricultural workers in the United States are foreign-born, with an increase of up to 1 percent per year who are undocumented (Reeves et al., 2002). While immigrants are not the only people working in U.S. agriculture, it is well documented that they encounter numerous systemic injustices. Among these injustices are labor rights violations, exposure to pesticides, poor working and living conditions, wage abuse, all of which relate to the core problem for immigrants working in U.S. agriculture—being voiceless. Without the right to organize, facing language barriers, and fear of retribution, immigrant farmworkers are left with little hope for a better future.

Organic agriculture

Using the definitions serving as a framework for this book, laborers in organic agriculture are also a population marginalized within the larger political-economic landscape of agriculture in the United States. Some of the injustices they face differ from those in conventional agriculture, and can differ across the scale of agriculture itself. In small-scale organic agriculture, there are a number of inequalities and challenges that overlap with those in conventional agriculture including, but not limited to, those involving immigration issues, substandard housing, lack of benefits and appropriate healthcare, and physically demanding and unsustainable working conditions.

As Thomas Arcury, Melinda Wiggins, and Sara Quandt point out, "The housing available to farmworkers is largely substandard and exposes workers and their family members to environmental health risks … In the eastern US, environmental and occupational regulations provide little protection to farmworkers" (2009, pp. 222–223). And, although much of the research focused on the environmental justice issues impacting farmworkers is from conventional agriculture and therefore heavily focused on pesticide use and related illness, it is still the case in organic agriculture that

> Although farmworkers experience high rates of occupational and environmental injury and illness, few programs and regulations have been designed to help reduce these outcomes. Farmworkers and their families in the eastern US seldom have health insurance, and many of them have

> limited access to health care. The few efforts to reduce farmworker injury and illness seldom consider the culture and educational attainment of farmworkers or the effects of a migratory lifestyle. Long-term consequences of occupational and environmental exposures are virtually unknown.
>
> (ibid., p. 223)

Examining the principles that serve as the foundation for environmental justice as a movement and a field of study, it becomes clear that organic agriculture itself is an effective place where we might invest resources as a solution to inclusion and equity. When considering that, "the movement for environmental justice is also about creating clean jobs, building a sustainable economy, guaranteeing safe and affordable housing, and achieving racial and social justice" (Cole & Foster, 2001, p. 17), it becomes particularly troubling (although perhaps not surprising) that the broader political economic landscape does not allow for justice within the very structures that represent the ideal for all of society, such as organic agriculture. As Green, Green, and Kleiner (2011) state, "the power that groups have over their own livelihood systems results, at least partially, from their position in the socioeconomic structure," and this is as true within organic agriculture as in conventional agriculture for those who labor in our fields. Due to large-scale, corporate, mechanized production, there has been a decline in the ability of small- and medium-sized family farmers to make a living, let alone provide truly just working environments for those who work on their farms (ibid.).

References

Agyeman, J. (2005). *Sustainable communities and the challenge of environmental justice.* New York: New York University Press.

Arcury, T. A., Wiggins, M. F., & Quandt, S. A. (2009). Conclusions: An Agenda for Farmworker Social Justice in the Eastern United States. In S. A. Quandt, & T. A. Arcury (Eds.), *Latino farmworkers in the Eastern United States* (pp. 221–233). New York: Springer. Available at: www.springerlink.com/content/k413847 031p03243/abstract/

Arnstein, S. R. (1969). A ladder of citizen participation. *Journal of the American Institute of Planners*, 35(4), 216–224. doi:10.1080/01944366908977225

Barry, B. (1989). *Theories of justice.* Berkeley, CA: University of California Press.

Benford, R. (2005). The Half-life of the Environmental Justice Farmer: Innovation, Diffusion, and Stagnation. In D. Pellow, & R. Brulle (Eds.), *Power, justice, and the environment: a critical appraisal of the environmental justice movement* (pp. 37–53). Cambridge, MA: MIT Press.

Bryant, B. (1995). *Environmental justice: issues, policies, and solutions.* Washington, DC: Island Press.

Bryant, B., & Mohai, P. (1992). *Race and the incidence of environmental hazards: a time for discourse.* Boulder, CO: Westview Press.

Callewaert, J. (2002). The Importance of Local History for Understanding and Addressing Environmental Injustice. *Local Environment*, 7(3), 257–267. Available at: http://doi.org/10.1080/1354983022000001633

Čapek, S. M. (1993). The "Environmental Justice" Frame: A Conceptual Discussion and an Application. *Social Problems*, 40(1), 5–24. Available at: http://doi.org/10.2307/3097023

CDC–NIOSH (2005, October) *Pesticide-related illness and injury surveillance: a how-to guide for state-based programs (2006–102).* Available at: www.cdc.gov/niosh/docs/2006-102/ (accessed January 21, 2012).

Chambers, R. (1997). *Whose reality counts?: Putting the first last.* London: Intermediate Technology.

Cole, L. W., & Foster, S. R. (2001). *From the ground up: environmental racism and the rise of the environmental justice movement.* New York: New York University Press.

Couto, R. (1999). *Making democracy work better: mediating structures, social capital, and the democratic prospect.* Chapel Hill, NC: University of North Carolina Press.

Couto, R. (2002). *To give their gifts: health, community, and democracy* (1st ed.). Nashville, TN: Vanderbilt University Press.

Delegates to the First National People of Color Environmental Leadership Summit. (1991). *Principles of environmental justice.* Washington, DC: First National People of Color Environmental Leadership Summit. Available at: www.ejnet.org/ej/principles.html

Di Chiro, G. (2008). Living Environmentalisms: Coalition Politics, Social Reproduction, and Environmental Justice. *Environmental Politics*, 17(2), 276–298. Available at: http://doi.org/10.1080/09644010801936230

Dobson, A. (1998). *Justice and the environment: conceptions of environmental sustainability and theories of distributive justice.* Oxford: Oxford University Press.

Dobson, A. (2003). Social Justice and Environmental Sustainability: Ne'er the Twain Shall Meet? In J. Agyeman, R. Bullard, & B. Evans (Eds.), *Just sustainabilities: development in an unequal world* (pp. 83–95). Cambridge, MA: MIT Press.

Faber, D. (2008). *Capitalizing on environmental injustice: the polluter-industrial complex in the age of globalization.* Lanham, MD: Rowman & Littlefield.

Faber, D., & McCarthy, D. (2003). Neo-liberalism, Globalization, and the Struggle for Ecological Democracy: Linking Sustainability and Environmental Justice. In J. Agyeman, R. Bullard, & B. Evans (Eds.), *Just sustainabilities: development in an unequal world* (pp. 38–63). London: Earthscan.

Foreman, C. H. (2011). *The promise and peril of environmental justice.* Washington, DC: Brookings Institution Press.

Freire, P. (2000). *Pedagogy of the oppressed* (30th anniversary ed.). New York: Continuum.

Gottlieb, R. (1993). *Forcing the spring: the transformation of the American environmental movement.* Washington, DC: Island Press.

Green, G., & Haines, A. (2001). *Asset building and community development.* London: SAGE.

Green, J. J., Green, E. M., & Kleiner, A. M. (2011). From the Past to the Present: Agricultural Development and Black Farmers in the American South. In A. H. Alkon, & J. Agyeman (Eds.), *Cultivating food justice: race, class, and sustainability* (pp. 47–64). Cambridge, MA: MIT Press.

Harsanyi, J. C. (1985). Rule Utilitarianism, Equality, and Justice. *Social Philosophy and Policy*, 2(2), 115–127. Available at: http://doi.org/10.1017/S026505250000323X

Heynen, N., Perkins, H. A., & Roy, P. (2006). The Political Ecology of Uneven Urban Green Space: The Impact of Political Economy on Race and Ethnicity in Producing Environmental Inequality in Milwaukee. *Urban Affairs Review*, 42(1), 3–25. Available at: http://doi.org/10.1177/1078087406290729

Holifield, R., Porter, M., & Walker, G. (2009). Introduction to Spaces of Environmental Justice: Frameworks for Critical Engagement. *Antipode*, 41(4), 591–612.

Holifield, R., Porter, M., & Walker, G. (2010). *Spaces of environmental justice*. Chichester: Wiley-Blackwell. Available at: http://search.ebscohost.com/login.aspx?direct=true&scope=site&db=nlebk&db=nlabk&AN=319179Honneth, A. (1995). *The struggle for recognition: the moral grammar of social conflicts*. Cambridge: Polity Press.

Hunold, C., & Young, I. M. (1998). Justice, Democracy, and Hazardous Siting. *Political Studies*, 46(1), 82–95. Available at: http://doi.org/10.1111/1467-9248.00131

Ife, J., & Tesoriero, F. (2002). *Community development: community-based alternatives in an age of globalisation* (2nd ed.). Frenchs Forest, NSW: Pearson Education.

Johnson, A. G. (2006). *Privilege, power, and difference*. Boston: McGraw-Hill.

Johnstone, G. (2007). *Handbook of restorative justice*. Portland, OR: Willan.

Morris, R. (2000). *Stories of transformative justice*. Toronto: Canadian Scholars' Press.

Nussbaum, M. C. (2011). *Creating capabilities: the human development approach*. Cambridge, MA: Belknap Press of Harvard University Press.

Obach, B. K. (2004). *Labor and the environmental movement: the quest for common ground*. Cambridge, MA: MIT Press. Available at: http://search.ebscohost.com/login.aspx?direct=true&scope=site&db=nlebk&db=nlabk&AN=122537

Pellow, D. (2000). Environmental Inequality Formation: Toward a Theory of Environmental Injustice. *American Behavioral Scientist*, 43(4), 581–601. Available at: http://doi.org/10.1177/00027642000043004004

Pellow, D., & Brulle, R. J. (2005). *Power, justice, and the environment: a critical appraisal of the environmental justice movement*. Cambridge, MA: MIT Press.

Phillips, R., & Pittman, R. H. (2009). *An introduction to community development*. London: Routledge.

Rawls, J. (1971). *A theory of justice*. Cambridge, MA: Belknap Press of Harvard University Press.

Reeves, M., Katten, A., & Guzman, M. (2002). *Fields of poison* (pp. 1–38). California: Pesticide Action Network, California Rural Legal Assistance Foundation, and United Farm Workers of America. Available at: www.panna.org/sites/default/files/FieldsofPoison2002Eng.pdf

Roht-Arriaza, N., & Mariezcurrena, J. (2006). *Transitional justice in the twenty-first century: beyond truth versus justice*. Cambridge: Cambridge University Press.

Rose, F. (2000). *Coalitions across the class divide: lessons from the labor, peace, and environmental movements*. Ithaca, NY: Cornell University Press.

Schlosberg, D. (2003). The Justice of Environmental Justice: Reconciling Equity, Recognition, and Participation in a Social Movement. In A. Light, & A. De-Shalit (Eds.), *Moral and political reasoning in environmental practice* (pp. 77–106). Cambridge, MA: MIT Press.

Schlosberg, D. (2004). Reconceiving Environmental Justice: Global Movements and Political Theories. *Environmental Politics*, 13(3), 517–540. Available at: http://doi.org/10.1080/0964401042000229025

Schlosberg, D. (2007). *Defining environmental justice: theories, movements, and nature*. Oxford: Oxford University Press.

Sen, A. (2009). *The idea of justice*. Cambridge, MA: Belknap Press of Harvard University Press.

Sologaistoa, E. (2011). *Farmworkers in the Southeast: Alabama, Florida, Georgia, Mississippi* (pp. 1–30). Tallahassee, FL: Florida Association of Community Health Centers.

SRIC: Voices from the Earth– 2003. (2003). Available at: www.sric.org/voices/2003/v4n1/ (accessed June 26, 2016).

Stephens, C., Bullock, S., & Scott, A. (2001). *Environmental justice: rights and means to a healthy environment for all* (Special Briefing No. 7). London: ESRC Global Environmental Change Programme. Available at: www.foe.co.uk/sites/default/files/downloads/environmental_justice.pdf

Sze, J., & London, J. K. (2008). Environmental Justice at the Crossroads. *Sociology Compass*, 2(4), 1331–1354. Available at: http://doi.org/10.1111/j.1751-9020.2008.00131.x

Taylor, D. E. (2000). The Rise of the Environmental Justice Paradigm: Injustice Framing and the Social Construction of Environmental Discourses. *American Behavioral Scientist*, 43(4), 508–580. Available at: http://doi.org/10.1177/0002764200043004003

The Black Commentator (n.d.). Environmental Justice for People of Color. *The Black Commentator*, 16. Available at: www.blackcommentator.com/16_re_print_pr.html (accessed June 26, 2016).

Walker, G. P. (2012). *Environmental justice: concepts, evidence, and politics*. London: Routledge.

Walzer, M. (1983). *Spheres of justice: a defense of pluralism and equality*. New York: Basic Books.

Watts, M., & Peet, R. (2004). Liberating Political Ecology. In R. Peet, & M. Watts (Eds.), *Liberation ecologies: environment, development and social movements* (pp. 1–43). London: Routledge.

Wenz, P. S. (1988). *Environmental justice*. Albany, NY: State University of New York Press.

Young, I. M. (1990). *Justice and the politics of difference*. Princeton, NJ: Princeton University Press.

2 Justice issues facing laborers in agriculture

In this chapter, I aim to paint a fuller picture of farm labor across the continuum presented in the Introduction of this book (see Figure 0.1) as an environmental justice issue specifically focusing on process claims/social dimensions of "who?" and "why?" ("How did this come to be?" will be discussed more thoroughly in Chapter 3 and the evidence claims/environmental and structural dimensions of "what?" and "where?" with justice claims are the focus of Chapter 6). In doing this, I acknowledge the contributions of many academics and activists regarding farm labor and farmworker advocacy and research, seeking to sew together their insights within the frame of environmental justice for the purpose of providing another avenue for understanding and change. In addition, the journey that our food takes from the fields to our tables is a complex one, and I acknowledge that this exploration only represents one dimension of the people who labor within that system, those who are on our farms, working in crop agriculture specifically.

Process claims: the social dimensions of farm labor

Who works in agriculture in the United States?

To understand more fully the dynamics at play within farmwork, we must first get a sense of who is working in U.S. agriculture. For this, I rely on publicly available data collected, analyzed, and published by several entities, including the U.S. Department of Agriculture (USDA) (the 2012 Census of Agriculture, the 2014 Census of Agriculture, the Organic Survey), the Economic Research Service (ERS), and the National Agricultural Statistics Service (NASS), the National Center for Farmworker Health, and the U.S. Department of Labor (National Agricultural Workers Survey (NAWS)). These data sources show how many agricultural workers there are, what types of crops they farm, the payment and benefits they receive, and their demographic characteristics, among other things.

There are many terms and definitions to consider when counting who works in agriculture in the United States, and, as seen in the data that follows, those terms and definitions influence the final numbers and results in any given report. In addition, depending on the definitions used when making data collection and analysis decisions, information about farmworkers' demographic characteristics, such as education level, wages and benefits, country of origin, citizenship status, and so on, can differ widely. Therefore, in what follows, I present a summary of all data to provide the fullest picture possible of labor in agriculture in the United States.

According to Charles Thompson (2002) in his Introduction to *The Human Cost of Food: Farmworkers' Lives, Labor, and Advocacy*, "Farmworkers are laborers who cultivate, harvest, and prepare a variety of seasonal crops for market or storage, including fruits and nuts, vegetables, horticulture, and field crops." Thompson points out that there are three categories of farmworkers which are not mutually exclusive, meaning an individual may fall in all of these during the course of a single season. These groups are (1) migrant farmworkers; (2) seasonal farmworkers; and (3) guestworkers. Migrant farmworkers travel more than 75 miles to obtain work in agriculture and are either shuttle migrants who travel back and forth between two primary locations during the year or migrants who follow the crop and move year round. Seasonal farmworkers live in one location, and are employed only during the growing season, which varies widely, depending on the area of the country. Guestworkers are in the country on a temporary visa called H-2A, which allows them to work in agriculture in the United States (ibid.). All of these categories of workers share both overlapping and unique struggles, which will be discussed in this book.

The last complete report from the National Agricultural Workers Survey (NAWS) describes the demographic and employment characteristics of hired crop farmworkers. The information in the report was collected from the U.S. Department of Labor's NAWS subcommittee between October 1, 2000 and September 30, 2002, through face-to-face interviews with 6,472 crop farm workers. It is important to note that there are several key limitations to the data that NAWS collects: (1) it only includes crop workers, so excludes workers in animal production, dairy, or other subsets of agriculture; (2) it excludes H-2A temporary agricultural workers; and (3) the survey occurs in workplaces rather than homes (Farmworker Justice & The National Center for Farmworker Health, 2015). In the study published by Carroll et al. in 2005, they found that 75 percent of all workers were born in Mexico, and 53 percent of the respondents were not legally authorized to work in the United States. In addition, they found that the average age of farmworkers was 33, with half being under 31, and 79 percent were men. Seventy-two percent of participants had had only one farm employer over the previous twelve months, and were therefore not migrating with the seasons for their work, although 60 percent reported that their current job was seasonal. Twenty-three percent of the workers reported having health insurance, but it was a benefit provided by the farm employer for only 8–12 percent of the workers. Seventy-nine percent of the workers reported being paid by the hour, 16 percent by the piece, 3 percent by a combination of hourly and piece rate pay, and 2 percent by salary, while 30 percent had family incomes below the poverty guidelines (ibid.).

In their presentation entitled "Changing Characteristics of U.S. Farm Workers: 21 Years of Findings from the National Agricultural Workers Survey" (Carroll et al., 2011), Daniel Carroll, Annie Georges, and Russell Saltz provide some updates to this earlier report, as well as summarizing trends over time. They point out that the average age of crop workers has increased in recent years to 36 years old (at its lowest in the history of NAWS, it was at 31 years old from 1992–2000). In addition, they point out that the percentage of Mexican-born farmworkers peaked in 1998–2000 at 79 percent, and also in that time the number of workers who migrate has decreased while increasing numbers of workers are naturalized citizens. According to the National Center for Farmworker Health, Inc. (2015), 72 percent of all farmworkers are foreign-born, with 68 percent of all farmworkers born in Mexico specifically. Also, 72 percent of workers were male, based on the 2011–2012 NAWS results. Those who live as migrant workers are still predominantly foreign-born and international (versus domestic). However, the percentage of unauthorized workers has remained fairly consistent, at around half of all farmworkers in the United States. Only 30 percent of all farmworkers report speaking English well, with the other 70 percent reporting "not at all" (35 percent), "a little" (27 percent), and "somewhat" (8 percent). In 2009, the average highest grade completed of all farmworkers in the United States was eighth grade.

In 2009, 50 percent of farmworkers had spent 9 or fewer years in farm work, and 81 percent reported spending all of their time on only one farm each year (in 1989 when NAWS began, only 52 percent reported only one farm employer per year), with the average number of weeks worked at 35 per year, compared to a low in 1995–1997 of 24 (Carroll et al., 2011). Finally, in this updated comparison, they looked at earnings and public assistance, and found that while average hourly earning had increased, those increases had not kept pace with the earnings of non-farm workers. While the percentage of farmworkers living below the poverty level had decreased, the share of workers receiving public assistance had increased by 30 percent since 1995–1997 (ibid.).

In addition to the National Agricultural Workers Survey, the United States Department of Agriculture (USDA) regularly conducts a Census of Agriculture through its National Agricultural Statistics Service (NASS). The most recent published report of this data is the 2012 Census of Agriculture. In 2012, there were over 2.1 million farms in the United States comprising 914,527,617 acres. The 2012 Census of Agriculture counted 566,426 of those farms (26.9 percent of the total) as employing 2,736,417 hired workers in the United States, and of the 2.7 million workers, over one-third worked for more than 150 days a year on the farm (36.5 percent). Migrant worker numbers on farms with hired labor and those that reported only using contract labor accounted for 436,570 workers, or 16 percent of the total number of hired farm laborers. One of the new items in the 2012 census was the number of unpaid workers, which comprise "agricultural workers not on the payroll who performed activities or work on a farm or ranch" (USDA, National Agricultural Statistics Service, 2014, Appendix B, B–24). The number of reported unpaid workers was 2,062,011 with

892,842 (42.3 percent of the total) farms utilizing unpaid workers. Additionally, in the latest census, expenses for hired farm labor as a percentage of total farm production expenses were 8.2 percent (they were 9.1 percent in the 2007 Census and 10.7 percent in the 2002 Census) with expenses for contract labor as a percentage of the total at 2 percent. Nationally, 88 percent of all farms fall under the USDA small farm definition, which are places that sell less than $250,000 in agricultural products annually (ibid.).

Along with the census and in conjunction with the USDA's Risk Management Agency, the USDA and NASS conducted an Organic Production Survey in 2008, 2011, and 2014.[1] The 2008 Organic Production Survey counted 14,540 organic farms and ranches in the United States, comprising 4.1 million acres of land. Of those farms, 10,903 were USDA-certified and 3,637 were exempt from certification. In addition, in 2008 they found that most producers sold their organic products locally. More than 44 percent reported selling within 100 miles of the farm, while 30 percent reported selling regionally (more than 100 miles but less than 500 miles from the farm). Another 24 percent reported selling nationally (500 or more miles from the farm), while only 2 percent reported selling internationally (USDA, National Agricultural Statistics Service, 2008). The 2011 Organic Production Survey collected information only from certified operations, whereas the 2008 and 2014 surveys collected information from certified, exempt, and transitioning operations. The 2011 survey found that Wisconsin led the nation with more than 110,000 acres harvested in 2011. Wisconsin was followed by New York, with organic growers harvesting more than 97,000 acres. California closely followed with more than 91,000 acres of organic field crops harvested in 2011 (USDA, National Agricultural Statistics Service, 2012).

By the 2014 Organic Survey, there had been a bit of a shift in the landscape of organic agriculture in the United States. The 2012 Census of Agriculture Organic Production Survey (2014) counted 14,048 organic farms and ranches in the United States (less than 1 percent of the nation's total farms, according to the 2012 Census of Agriculture total of 2,109,303 farms), comprising 3,670,560 acres of land (4 percent of the nation's farm acreage of 914,527,617 according to the 2012 Census of Agriculture).[2] Of those farms, 12,595 were USDA-certified and 1,453 were exempt from certification. The 2014 Organic Production Survey found that California now leads the nation with more than 687,000 acres harvested on certified or exempt farms. California is followed by Montana, with organic growers harvesting more than 317,000 acres. Wisconsin, Oregon, and New York follow with more than 200,000 acres of organic field crops harvested each. Some 8,439 of these farms (60 percent) report using hired agricultural labor, including contract labor (USDA, National Agricultural Statistics Service, 2015).

The Economic Research Service (ERS) of the U.S. Department of Agriculture uses data from the Agricultural Resource Management Survey (ARMS) to examine the economic well-being of farm households in the United States. The ARMS is a nationally representative survey that targets around 5,000 fields and 30,000 farms each year to gather information on their financial condition, production practices, and economic well-being (USDA, n.d.a). Data are analyzed on the income generated by on-farm and off-farm sources, wealth as measures as a sum of all farm and nonfarm assets minus debt, and health insurance. In addition, ERS analyzes information about farmworkers, acknowledging that:

> Wages, salaries, and contract labor expenses represent roughly 17 percent of total variable farm costs, and as much as 40 percent of costs in labor-intensive crops such as fruits, vegetables, and nursery products. Hired farmworkers continue to be one of the most economically disadvantaged groups in the United States.
>
> (USDA, n.d.b)

Using the Farm Labor Survey (FLS) of the National Agricultural Statistics Service (NASS), the ERS examined data from just over 1 million (1,063,000) hired farmworkers in the United States. Their data shows that hired farmworkers work primarily on large farms with sales of over $500,000 per year, and that they make up only a third of all the people working on farms, with the other two-thirds being the farm operators and their families. Some 56 percent of farmworkers work in crop agriculture, while the remaining 44 percent work in livestock agriculture. The largest regional concentrations of farmworkers are in the Southwest and the Midwest, with California and Texas accounting for one-third of total farmworkers. In analyzing the demographic characteristics of hired farmworkers, they divide them into two groups, the first being laborers and field supervisors, and the second being hired farm managers. For the first group (laborers and field supervisors), half were Hispanic, 31 percent had less than a ninth grade education, 59 percent were U.S. citizens, and were overall more likely to be young and unmarried. Among those in the second group, mostly non-Hispanic whites, 6 percent had less than a ninth grade education, and 91 percent were U.S. citizens. The hourly wage earnings for farmworkers in nonsupervisory positions was reported at $10.80 in 2012 (USDA, n.d.b).

Why? The conditions contributing to lack of justice for farmworkers

Charles Thompson states, "At its very foundation, the agricultural system in its present state depends on farmworkers. While the arrangement is not a holdover from the days of slavery, comparisons of present-day farm work to slavery are warranted" (2002, p. 10). There are many books and articles dedicated to the exploration of the injustices that farmworkers face in the United States that back Thompson's claim, and many times make it feel like one is reading about pre-Civil War conditions rather than those in modern-day farming. Farmworkers are seen as an economic input or cost of production in farming, and as land has been privatized in the United States, it has separated those involved in agriculture into two classes: owners and workers (Allen, 2004). In her book, *Together at the Table*, Patricia Allen discusses the dynamics of those two groups as being wholly different due to the fact that laborers only have their own labor to sell, while the owners or farmers, are entrepreneurial, property-owning, and hire the workers. In addition, she points out that the common assertion that family farms are more ideal for laborers than are large, corporate farms is a fallacy due to these two classes and the power dynamics at play, no matter the scale of agriculture. These dynamics inarguably continue into agriculture today.

Allen also describes two levels of political ecology involved: the local and the structural. At the structural level one must consider the larger socioeconomic and historical dimensions that cause problems that manifest locally. In the case of agriculture, a number of things have kept a steady supply of workers in the system, therefore keeping the cost (i.e. wages for workers) low. For these reasons, forces for change within agriculture have historically represented farmers' interests rather than those of farmworkers, meaning that this cycle perpetuates over time.

Toward this end, David Rothenberg points out that the grower community has unified time and again around the goal of keeping labor costs low. In doing so, they "have fought legislation designed to increase wages, improve safety conditions and regulations, provide field sanitation, regulate pesticides and herbicides, and provide protections for farmworker organizing" (1998, p. 76). For his (1998) book, entitled *With These Hands: The Hidden World of Migrant Farmworkers Today*, Rothenberg spent years collecting information from workers, contractors, growers, lobbyists, government investigators, union organizers, and other stakeholder groups such as physicians and teachers. Through all of this work, he came to the conclusion that the marginalization of farmworkers is a reflection of the inequity of the system itself. Further, he points out that because of a short period of time in which to harvest most crops in many areas, "the agricultural industry … has two options – either create incentives to ensure that workers will accept difficult, temporary jobs, or find workers who have few other options" (ibid., p. 72).

All of this results, as Margaret Gray (2014) points out, in a system of paternalism within agriculture between farm owners and operators and farmworkers. Paternalism can be even more of a danger on small and family-scale farms due to the closeness of the relationships, the necessity for close monitoring due to a reliance on hand labor, and a lack of middle management. It is clear that paternalism abounds in both intentional and unintentional ways when it comes to laborers in the agricultural system. Farmworkers' reliance on their employer(s) for their and their families' housing, safety, and employment when many are already vulnerable due to their socioeconomic status, lack of education, and poor English-speaking background, tenuous citizenship status and/or being undocumented, and so on leads to a recipe for failure when it comes to justice for farmworkers. It is important to note that the role of policy over time has also created the context within which these injustices are rationalized and supported by law. In what follows, I will discuss how this lack of justice manifests itself in these farmworkers' lives, and in Chapter 3, I will provide a brief history and record of policy pertaining to the injustices faced by this group of laborers.

Evidence claims: the environmental and structural dimensions of farm labor

What issues are farmworkers facing?

There are a multitude of issues that affect farmworkers in the United States. As Brian Obach (2015) points out in his book, *Organic Struggle: The Movement for Sustainable Agriculture in the United States*, the history of agricultural labor is one derived from slavery, indentured servitude, and the exploitation of immigrants. In Jill Lindsey Harrison's (2011) *Pesticide Drift and the Pursuit of Environmental Justice*, she details the scope of concerns studied by scholars, ranging from economic justice for small-scale farmers, food justice issues facing low-income eaters, and the labor issues faced by farmworkers. It is important to note, as she further points out, that the benefit to the farm owner in violating labor and environmental regulations is greater than the risk of possible fines and other sanctions if discovered. Because the farm labor market is oversaturated, workers are unlikely to question the unfair labor practices so as not to risk the possibility of losing employment. These dynamics create a context within which marginalization and injustice can easily occur. These injustices include occupational injury and illness (including pesticide-related illness), lack of access to healthcare, housing and hunger, issues pertaining to immigrant and migrant workers, issues unique to women and children working in agriculture, and low wages, lack of stable employment, and the inability to organize. It is important to note that while they are separated here into sections for the sake of more fully exploring each issue, these injustices are not mutually exclusive, and most often relate to and compound one another.

Occupational injury and illness and the lack of access to healthcare

In *Latino Farmworkers in the Eastern United States*, John May (2009) provides a staggering list of the different occupational injuries and illnesses faced by farmworkers. This list includes ergonomic problems, health problems, eye injury, hearing loss, heat stress, musculoskeletal injuries, severe acute injury and fatality, skin disease, and transportation fatalities. Further, he relates many of these issues to exposure to pesticides and comments on the exacerbation of the symptoms due to lack of access to adequate, or any, healthcare. Allen (2004) also points out the health problems that farmworkers face, including arthritis, cancer, respiratory disorders, and injuries from machinery, as well as poisoning, reproductive problems, and death caused by pesticide exposure. In reviewing studies related to farmworker health, the National Center for Farmworker Health makes the following statement, illustrative of the interrelation of health with so many other issues faced by farmworkers:

> [P]overty, frequent mobility, low literacy, language and cultural barriers impede farmworkers' access to social services and cost effective primary

> health care. The small percentage of farmworkers who do take advantage of health services are faced with further issues: limited means of transportation, prejudice because of their status as migrants, the lack of time-efficient healthcare delivery methods and the medical referral system.
>
> (National Center for Farmworker Health, 2012, p. 3)

Seth Holmes (2013) studied the social and economic structures that impact healthcare for farmworkers, presented in his book *Fresh Fruit, Broken Bodies: Migrant Farmworkers in the United States*. He conducted his research when he was training as an anthropologist and medical doctor, and during this study he heard from workers time and again that doctors did not know how to help them. Given that evidence points to farmworkers having one of the highest incidences of workplace fatality and disabling injuries of any group of laborers, the lack of adequate healthcare is troubling (Rothenberg, 1998). Despite these grave dangers faced daily by farmworkers, the National Center for Farmworker Health (National Center for Farmworker Health, 2012, p. 2) asserts, "Farmworkers rarely have access to workers' compensation, occupational rehabilitation, or disability compensation benefits. Because workers' compensation benefits are state-dependent, agricultural workers are often further challenged by the qualifications and requirements of each individual state." Additionally, health insurance access for farmworkers presents its own set of difficulties. Agriculture is one of the most dangerous sectors for workplace injuries, and H-2A workers legally must be insured in the United States. Yet expenses associated with insurance, long working days, and billing processes that occur via mail make even this step towards healthcare access logistically tricky. Many advocacy organizations focus on enrolling these H-2A workers and assisting in the process. This still leaves a majority of undocumented workers in the United States without health insurance, and the transience of the work is not covered (many state insurance policies do not transfer from state to state). Even with insurance, the time and transportation required for adequate medical attention may be a sacrifice farmworkers are not willing to take (Cancino, 2016).

Pesticide-related illness

One of the biggest occupational hazards faced by farmworkers is exposure to pesticides, and the resulting illness and death. Not only are pesticides extremely dangerous, but often the symptoms of illness related to them also lurk as other things, such as a cold or the flu; long-term and chronic disease caused by them is almost impossible to trace to the source. A number of studies have been conducted over time, providing not only empirical evidence that pesticides are harmful and which ones are most dangerous, but also how farmworker safety is tied to them not only through exposure, but through education about them, signage and warnings pertaining to them, access to proper protective equipment, and processes through which workers can raise and address concerns. In what follows, I provide a brief survey of studies that provide context and scope to this particular issue.

In *Fields of Poison* (Reeves et al., 2002), the Pesticide Action Network North America (PANNA) reports on the use of pesticides and regulation lapses in the state of California, for the purposes of making a number of recommendations to the Department of Pesticide Regulation. This report, published in 2002, has become a model for other states and regions as they try to advocate for an improvement in the working conditions for farmworkers and highlight the harmful effects of pesticides to consumers. By highlighting the types of pesticides most frequently used and causing harm, the crops that use these poisons, and the data regarding regulation oversight, PANNA offered suggestions focusing on an "urgent need for safer agriculture and better worker protections" (ibid., p. 7). Chief among these nine recommendations are protestations about the importance of reporting, regulation, and enforcement. They provide a methodology for collecting and interpreting the data, and this structure has been used by numerous other entities as a gold standard in making a demonstrable point for change.

The PANNA report noted a yearly average of 475 cases of agricultural pesticide poisoning statewide from 1997 to 2000. They point out that this is likely much lower than the actual occurrences because many cases go unreported. The reasons these cases go unreported are nuanced and complex, and related to many of the same reasons that reporting other violations is also low. However, with pesticide-related illness, there are also unique reasons for underreporting, including "doctors' failure to recognize and/or report pesticide-related illnesses; failure of insurance companies to forward doctors' illness reports to the proper authorities; or farmworker reluctance to seek medical attention for suspected pesticide exposure" (ibid., p. 4). The short-term and ongoing symptoms of pesticide-related illness can include "headache, nausea, lip numbness, swollen lips and tongue, irritated throat, nose, and eyes, trembling, and momentary blackout" (ibid., p. 11) in addition to vomiting, itching, a rash, and difficulty breathing. Long-term health effects include cancer and birth defects, but it can be more difficult to trace long-term illness to the source or incidences of pesticide exposure.

Illness related to pesticides can be caused by pesticide drift or direct residue exposure, and employers are supposed to follow a number of safety precautions to ensure that their workers are safe from both types of exposure. Over one-third of farm employers violated safety regulations such as providing washing and decontamination facilities, provision and use of protective equipment, and information about the pesticides being used (ibid.). In a later study, Harrison (2011) gathered data from 2001–2009 through interviews, observation, and printed materials. During her research, she found a relationship between the lack of effective environmental problem solving and the inequalities and oppression experienced by farmworkers, particularly pertaining to pesticides and pesticide drift. These findings are discussed further in a later section in this chapter regarding justice issues faced by workers in conventional agriculture.

On the tail of recommendations like those put forward by PANNA, the Centers for Disease Control and Prevention's (CDC) National Institute for Occupational Safety and Health (NIOSH) published *Pesticide-Related Illness and*

Surveillance: A How-To Guide for State-Based Programs (CDC–NIOSH, 2005). While the CDC had been providing financial and technical support for the surveillance of pesticide use and illness since 1987, this report signaled a turning point for the provision of free and easily accessible resources for agencies interested in monitoring practices and assisting farmworkers. The comprehensive guide includes:

> (1) information about the importance of pesticide poisoning surveillance; (2) mechanisms to improve reporting of cases to surveillance programs; (3) methods to investigate reported cases; (4) guidance on using the case definition; and (5) additional resources on pesticide toxicology, pesticide usage, governmental partners, and surveillance.
>
> (ibid., p. iii)

The hopes portrayed in this guide are that farmworker organizations are able to take matters into their own hands and understand the intricacies of the legal processes surrounding government oversight and regulation, as well as how to navigate those various channels.

In their study "Pesticide Safety among Farmworkers: Perceived Risk and Perceived Control as Factors Reflecting Environmental Justice," Arcury, Quandt, and Russell (2002) interviewed 293 farmworkers, and used interview data to develop scales for perceived pesticide risk and perceived pesticide control. They then studied the relationship between those areas of risk and control with safety knowledge and safety-related behavior. They found that farmworkers who reported higher levels of perceived control were equipped with safety knowledge and information about behavior to reduce risk, such as warning signs and protective equipment. Perceived risk itself, when not coupled with perceived control, was not enough to influence behavior relating to safety around pesticides. Given these conclusions and their assertion that environmental justice necessitates that communities have control over their environment, they make the case that effective pesticide safety education must account for issues of farmworker control, that "environmental justice education programs must not be limited to information. Rather, these programs must help affected communities to gain control of the process for implementing change" (ibid., p. 239).

In their report entitled *Exposed and Ignored: How Pesticides Are Endangering Our Nation's Farmworkers*, Farmworker Justice states:

> There is an estimated 5.1 billion pounds of pesticides applied to crops each year, and thousands of farmworkers experience the effects of acute pesticide poisoning, including headaches, nausea, shortness of breath, or seizures. Pesticide exposure leads to chronic health problems, such as cancer, infertility (and other reproductive problems), neurological disorders, and respiratory conditions.
>
> (2013, p. 4)

After exploring the issues facing farmworkers due to pesticide exposure, they make a number of recommendations about how to reduce the risk of pesticides for these agricultural laborers. These recommendations include revisions to the Worker Protection Standard to increase education on pesticides, an understanding of the risks associated with, and preventative measures regarding pesticide use; Spanish translations of pesticide labels, the implementation of buffer zones in residential areas and around schools, reporting of pesticide use and poisoning to the Environmental Protection Agency (EPA), and increased funding for research on the health impact of pesticide exposure and technology to prevent exposure (ibid.).

Housing and hunger

While pesticide exposure is one of the greatest and most studied risks faced by farmworkers, housing and hunger are also justice-related issues associated with their status as a marginalized and often poor population as well. Living in employer-provided housing is often a condition of employment. The accessibility of this housing is rarely, if ever, inspected or held to code. Christopher Holden (2002) frames the argument around problems farmworkers face with housing as one exacerbated by lack of actual data documenting the conditions faced by farmworkers. He argues that policy work is not possible without this data. Then he mentions an unpublished study done during the Carter administration to document farmworker housing conditions, during which researchers found that migrant housing was more likely than year-round housing to be substandard, measured by things such as lack of heat, plumbing, and laundry facilities, overcrowding, and structural damage and deterioration. In an attempt to provide data, the Housing Assistance Council, in partnership with Farmworker Health Services, Inc., embarked on a study beginning in 1997 to survey farmworker housing conditions, particularly in migrant streams. Holden details the results of the Eastern stream for this chapter. While a lot of information was uncovered to support the predominant sentiment and argument that the housing situation for farmworkers is dire, perhaps most notably it provided empirical data (rather than journalistic accounts) of the state of affairs in relation to farmworker housing. It reinforced the perceptions that the main issues are overcrowding, the high cost of housing in relation to income, the lack and/or disrepair of basic appliances and sanitary facilities, pesticide exposure in housing that borders fields, and dilapidated housing with external and/or internal structural damage that was oftentimes severe.

Compounding these issues are dimensions relating to citizenship status and worker eligibility (ibid.). While employers utilizing H-2A workers are required to provide housing and have that housing inspected, those who do not employ these guestworkers are more likely to risk the penalties or fees associated with non-compliance because inspections are few and far between. For undocumented workers (around 50 percent of the farmworkers in the United States), they "are not eligible to reside in housing developed by federal funds and except for emergency shelter are prohibited from living in housing developed through various states' programs" (ibid., p. 190). This is significant because of the interrelationship of issues associated with citizenship status, and the cycle of injustice within which undocumented workers fear retribution if reporting violations. Corroborating the findings of this study, Patricia Allen (2004) points out that at least one-third of farmworkers and farmworker families live in housing supplied by their employers. It is in this housing and during work hours that she reports that workers lack access to restrooms and fresh drinking water. In addition to this, they are one of the populations at greatest risk of hunger.

Sandy Brown and Christy Getz (2011) have spent years researching issues facing farmworkers in California, the state with the largest number of farm laborers in the United States. In their chapter in *Cultivating Food Justice: Race, Class, and Sustainability* they point out that those who produce our nation's food are the most likely to be hungry or food-insecure. In their research, around 50 percent of the farmworker homes they surveyed met the USDA definition of food-insecure, and that income was the strongest predictor of food insecurity. Another strong influence was the status of the household members' work authorization or citizenship because many times they were ineligible for public assistance. This lack of authorization to work legally in the United States creates barriers to accessing food and adequate housing through formal channels.

A recent study by Elizabeth Kiehne and Natasha S. Mendoza (2015) examined the prevalence, impact, risk factors, and coping strategies related to migrant and seasonal farmworker food insecurity. In their article, they point out that migrant and seasonal farmworkers (who are predominantly Latino) are 400 percent more likely to experience food insecurity than the general population in the United States. After conducting a thorough literature search, they found ten studies that address food insecurity and hunger among farmworkers. They point out that across these studies, food insecurity has a number of health effects, including birth defects, depression, diabetes, and obesity. Risk factors included households with children, those who migrate with the seasons, and lack of transportation and ability to prepare food. Lack of education was also found to contribute to an increased likelihood of food insecurity, and, as explored earlier, the average educational level of the farmworker population is low compared to that of the general population.

Migrant and immigrant issues

Earlier in this chapter, I noted that at least 72 percent of all farmworkers in the United States are foreign-born, and around 50 percent are undocumented workers. Given these large percentages, it stands to reason that there are a number of issues related to immigrant status and the necessity of migrating throughout the season to find eligible employment. While not mutually exclusive from others discussed in this chapter, these issues create additional conditions that are important to consider. Because both Holmes and Rothenberg conducted anthropological research, their books are helpful first-hand accounts of the types of obstacles immigrant and migrant laborers face. In his research, Holmes (2013) documented the dynamics and dangers of border crossing, including robbers, kidnappers, rattlesnakes, and the heat.

Rothenberg (1998) discussed in depth the debt peonage that many farmworkers have to contractors. Their debts often include payment needed for food, rent, loan payments, and credit purchases of wine, beer, liquor, and drugs. Many times this debt adds up to more than the farmworker makes per week, meaning that they owe money come payday despite working long hours in the fields. In his travels, he documents that migrant workers are immigrants, poor, and/or homeless, and are regularly persons of color, including black and Latino/a, which is corroborated by the statistics compiled by the various sources discussed earlier in this chapter. Being transient, or being constantly dis- and re-located, creates special needs for migrant farmworkers and their dependants, as they have perpetual outsider status in the communities in which they work and live during different parts of the season. Not only do they have to find housing and transportation in each new location, but they have to do so while remaining largely invisible to members of those communities.

While it is difficult to imagine living under this pressure and these dynamics, these very dynamics also make it difficult to study this large sector of the farmworker population to get a true sense of who they are and what they face. A report that came out in late 2011 and is worth highlighting is *Farmworkers in the Southeast* (Sologaistoa, 2011). It illuminated how difficult it can be to gather information about the farmworker population due to its migratory patterns, but also showed that it is possible to work together across state boundaries to talk about the issues facing farmworkers on a larger scale. It shows that by working together, we can sometimes gather more representative information about this mobile and shifting population. Sologaistoa structures the recommendations for various audiences, from government regulatory agencies to farmworker organizations. This report demonstrates the value of scaling up data collection and subsequent calls to action, suggesting that research coalitions are a valuable possibility to explore for the future. In fact, conducting research in this way may be the only possibility for getting an accurate sense of what these workers face.

While the study is outdated, George S. Rust (1990) conducted important research in the study, "Health Status of Migrant Farmworkers: A Literature Review and Commentary" where he went through studies from 1966–1989 regarding migrant families and the health of agricultural workers and farmers. His findings are in alignment with many of the justice issues discussed in this chapter (chronic disease, pesticide exposure, skin disease, etc.) but what he also offers is an explanation for why migrant populations are so difficult to learn more about, and why, despite that, it is also so important that we do so. He states:

> In migrant health we lack even such basic data as crude death rates, median survival, infant and maternal mortality, and incidence of permanent disability. One of the major difficulties in computing these is the lack of a precise denominator. We do not know how many migrants there are,

> because of the transient nature of the population, their migrations into and out of the US (both randomly and seasonally), undercounting of workers who meet the legal definition of a migrant but do not fit ethnic or demographic stereotypes, and the desire of many immigrant workers (documented or not) to avoid contact with government agencies.
>
> (ibid., p. 1214)

Because of this, he points out that immigrant and migrant farmworkers are undercounted by nearly every data source, and it is nearly impossible to get a sense of baseline morbidity and mortality rates. Therefore, one cannot make any assertions of whether things are getting better or worse over time.

To add to the confusion of change over time, David Griffith (2006) points out that H-2A workers do get fed up with their situations, and sometimes there is hope when they try to leave their employers. However, when they leave, they also risk not being called back to work should things go awry, many times this ensures compliance even when workers are aware of mistreatment. When it is successful, however, it can ensure farmworkers gain a better standard of living and the possibility of organizing more broadly to advocate for better working conditions. Toward this end, he states:

> Working off the contract exposes H-2 workers to more than simply other sectors of the economy, other employers, other jobs. Through the increased communication taking place among Latino co-workers, H-2s working off contract learn about housing opportunities, healthcare systems, places where Latinos gather, new churches, and centers that address issues of social justice.
>
> (ibid., p. 213)

Given that these workers make up such a significant proportion of the farmworkers population in the United States, getting a handle on the unique issues they face that are wrapped up with issues associated with lack of citizenship is essential.

Women and children farmworkers

Most recent data from NAWS indicates that 72 percent of farmworkers in the United States are male (Farmworker Justice & The National Center for Farmworker Health, 2015). The remaining 28 percent of workers are comprised of women and children. Clearly, both women and children working in agriculture face the same dangers and injustices as men, but there are added layers to the justice-related issues each group encounters. As Deborah Barndt, author of *Tangled Routes: Women, Work, and Globalization on the Tomato Trail* notes, "Job categories and divisions of tasks within tomato production have evolved over the decades (indeed centuries) to reflect and reinforce institutionalized classism, sexism, racism, and ageism" (2002, p. 230). While Barndt is commenting specifically upon the dynamics within farming tomatoes, it is fair to say that her notion applies throughout agriculture.

Hinda Seif (2008) spent several years doing ethnographic fieldwork to explore the lives of immigrant women farmworkers, with a specific focus on their powerlessness within the political systems that shape their lives:

> Because of their unenfranchised status, Mexican immigrant women are invisible to the quantitative methods deployed by political scientists such as analysis of election exit polls, voter registration lists, and electoral results. Women who lack lawful immigration status are especially marginalized in research on political activism and participation.
>
> (ibid., p. 79)

Seif points out that due to their roles not only as workers, but also as family care providers, women farmworkers face an extra burden of pressure. However, Seif notes that this creates an added incentive for women farmworkers to participate in organizing efforts rooted in community, and ultimately fighting for things that will not only benefit them, but their families as well. These women added to UFW strength, and the "infusion of women's leadership shaped creative organizing tactics that brought women like Marta and Fernanda into the union despite their marginal position in agribusiness" (ibid., p. 91). The women involved in Seif's study do note that the largest barriers to participation include language and potential retribution from employers, as she notes by stating, "The barriers to women farm workers' political engagement are magnified and qualitatively distinct because of their immigration status and agricultural setting" (ibid., p. 94).

Corresponding with their lack of voice through formalized political channels, studies also show that women farmworkers face increased risk of violence—both sexual violence at the workplace (the farm) and domestic violence. Joe Gorton and Nikki Van Hightower (1999) used data collected in 1994 by the Migrant Clinicians' Network regarding domestic violence among farmworker women, to examine both its prevalence in this community as well as predictive social and demographic factors. They found that 17.4 percent of respondents reported being physically or sexually abused in the previous year, at an average of 11.5 incidents over that time. The biggest predictor of physical or sexual abuse was found to be drug and/or alcohol use of husbands, boyfriends, or companions and the only thing that (in a statistically significant way) decreased the likelihood of abuse was pregnancy. The National Farm Worker Ministry and the Southern Poverty Law Center report that 90 percent of farmworker women have identified sexual harassment as a major problem (Southern Poverty Law Center, n.d., National Farm Worker Ministry, n.d.b). Their vulnerability is exacerbated by "their lack of familiarity with their legal rights, lack of access to service providers, lack of transportation, the extreme poverty in which they live and lack of formal education and, in some cases, English language skills" (Southern Poverty Law Center, n.d.).

In addition to these political and violence-related challenges, perhaps one of the better-explored areas of the effect of working in agriculture for women is around reproductive health. Pesticide exposure leads to a host of health issues including infertility and other reproductive issues (Farmworker Justice, 2013). In her piece on the reproductive health of farmworker women, Galarneau examined reproductive oppression, its social inequity causes and lived experiences of women, and strategies for change employed by farmworkers. She comments on the role of macro-level policy (trade and labor) in impacting reproductive oppression, stating, "It is important to recognize that these broad policies regulate and compromise women's income, work, migration, health, and safety, and thus are critical elements of a comprehensive understanding of reproductive oppression" (2013, pp. 147–148). She further explores different areas of reproductive oppression, including labor and occupational conditions, health care, and social relations all of which come together to "interact to create a context that regulates, controls, and exploits women farmworkers" (ibid., p. 151). She, much like Seif, finds that farmworker women exercise their voice through organizing to change the conditions in which these injustices occur.

To address the dangers that pregnant farmworker women face in their daily work, Runkle et al. (2014) turn to educational interventions designed to inform farmworker women of the dangers of heat exposure, stress due to physical labor, and exposure to pesticides on their developing fetuses and young babies and children. This education is important because previous studies have found that farmworker women have little knowledge of the threat posed to their unborn or young children's health, and when they do discover that there is a risk, they are unaware of or without channels to take action for their own protection (Flocks et al., 2011; Kelley et al., 2013). This creates an environment in which:

> Female farmworkers consistently believe that pesticide exposure can adversely affect pregnancy and child health, yet they do not receive information about these specific risks in mandated pesticide training nor, as our related work has shown, do health care providers routinely discuss these issues with them during pregnancy.
>
> (Flocks et al., 2011, p. 631)

The study unearthed a number of fertility-related issues faced by farmworkers, including infertility and male sterility, rashes in the female genital area, miscarriages, birth defects, developmental disabilities in children, and respiratory illness in children exposed in uterus.

In their follow-up study consisting of interviews with healthcare providers, Kelley et al. (2013) interviewed healthcare providers about their work with pregnant farmworker women, and found that the data collected fell into five broad themes: (1) information collected at intake; (2) barriers to healthcare; (3) patient occupation, culture, and pregnancy health; (4) occupational and environmental hazards during pregnancy; and (5) healthcare provider needs.

Their findings in these areas showed significant room for improvement, including lack of collection of occupation at intake, to turning patients away for lack of insurance and/or showing up for prenatal care too late in the pregnancy. These findings illuminate further injustice faced by farmworker women as they navigate their high-risk pregnancies.

In another study, Runkle et al. (2013) surveyed workers at nurseries and ferneries in Florida about their perception of risk due to pesticides, then evaluated the accuracy of these perceptions by testing urine samples of participant workers for the presence of pesticides. What they found was that while nursery workers perceived less pesticide-related risk than those in ferneries, the presence of pesticides in their urine was higher. The two pesticides they tested for have been linked through previous studies to increased risk of birth defects, infertility, and miscarriage. This series of important studies shows the relationship between occupational hazards and pesticide exposure to pregnant farmworker women, yet also illustrate that lack of knowledge and informed, accessible formal channels of reproductive treatment are unattainable.

For the estimated 300,000–800,000 agricultural workers under the age of 18, pesticide exposure and agricultural hazards have additional effects. Childhood health is particularly vulnerable to the presence of pesticides, with studies discovering associations with asthma and respiratory function, birth defects, brain and nervous system harms, cancer, and diabetes and obesity (Marquez & Schafer, 2016). Excluded from the regulations that protect children in other labor sectors, child farmworkers can work as young as 14 on larger farms and without age requirements on smaller farms. While not allowed to work during school hours, there are no limits on hours worked outside of those restrictions. The data on this demographic is unreliable due to the nature of many child workers registered under adult documentation. Enforcement for any regulations for children remains weak with little funding for inspection, compliance, or enforcement mechanisms (Bon Appétit Management Company, United Farm Workers, 2011).

Wages, lack of stable employment, and inability to organize

As a central tenet of a free market economy, individuals work in exchange for pay and benefits. There is also an assumption held in both principle and reflected in law and practice that these wages should be fair, and individual workers should have the opportunity to use a process when they are not. Complicated by many of the factors, farmworkers are left out of this system of fair remuneration and lack viable pathways through which they can report violations and/or hold violators accountable. In his 2008 Report for the Economic Research Service for the U.S. Department of Agriculture, Kandel (2008) reminds us that, “Hired farmworkers earn less than other workers. Median weekly earnings of full-time farmworkers are 59 percent of those for all wage and salary workers. Poverty among farmworkers is more than double that of all wage and salary employees.”

In 1998, farmworkers made $7500/year, migrant farmworkers make an average of $5000/year, and migrant farmworkers who rely on a contractor made only $3500/year. In May 2011, the Economic Research Service (USDA, n.d.b) reported a range of $20,000–$24,000/year as the salary of farmworkers. While this most recent data may be skewed higher than actual earnings, it is important to note that the median reported salary ($8.99/hour) is lower than any other worker type, both in farm and some nonfarm (housekeepers, construction, groundskeeping) occupations. The National Farmworker Ministry reported that as of 2009, farmworkers "work 42 hours per week and earn $7.25 per hour on average, but this average varies greatly" (National Farm Worker Ministry, n.d.a). They give a range of $10,000–$12,499/year for individual farmworkers and $15,000–$17,499/year for a farmworker family, pointing out that 30 percent of all farmworkers live below the poverty line. It is also important to note that many farmworkers are not paid hourly, but are instead paid by piece rate, wherein they are paid according to how much they pick during a given time period. While piece rate pay might result in getting paid more for fast or especially skilled laborers, it creates many issues resulting in workers taking less breaks, ultimately impacting their health and compounding the risk of occupational hazards as described previously. And, although there are laws in place that ensure that piece rate workers are paid the minimum wage, should they not hit it with the amount they pick, many farms are exempt from these regulations because they are small farms or because they employ workers hired by contractors.

Therefore, while the landscape has changed incrementally since the 1980s, the reality of substandard pay remains. And while there are relief programs that try to lessen farmworker poverty and vulnerability, many of these programs reinforce the inevitability of marginalization rather than working to change the system within which that marginalization occurs. Obach (2015) illuminates the fact that fair wages are hard to find anywhere in the agricultural system, including organic agriculture. In addition, he laments the shrinking numbers of farmworkers involved in unions beginning during the Reagan administration in the 1980s after the momentum gained by the United Farm Workers (UFW) and Cesar Chavez and Dolores Huerta in the 1960s and 1970s. By examining the public timeline made available by the United Farm Workers, it is clear that the 1980s were a time when this movement lost a lot of ground, and not only did many farmworkers lose their UFW contracts during this time, but faced blacklisting and violence for participating in the union and in union elections (UFW, n.d.).

While her research focuses on farms in the Hudson Valley, New York, Gray (2014) unpacks and explores the dimensions of paternalism and power disparities as significant contributors to labor control, tied intricately to questions of livelihood relating to aspects such as pay. In farmwork, there are large fluctuations in the hours and days that are worked from week to week, which lead to economic instability. In the Hudson Valley, the average farm size is 143 acres, and a high concentration of those are fruit, vegetable, and/or horticultural farms, which are all seasonal in that part of the country. Therefore, many of the jobs in Hudson Valley agriculture are seasonal in nature, filled with migrant workers who are more susceptible to poor wages and worse working conditions due to this lack of opportunity to organize and a fear of retribution if one were to take legal action over injustices. In addition to the seasonality of the work, due to the rural nature of many of these farms and limited options for transportation, workers are lonely and isolated. These factors contribute to Gray's assertion that there is a "psychology of extreme compliance and sacrifice that is shaped by their situation" (ibid., p. 47).

Gray's (2014) assertions here, in addition to those that point out that many workers' concerns are also integrally tied to insecurity in their citizenship statuses, are illustrative of the cycle of injustice for farmworkers that has repeated itself over time and throughout history. Further, Allen's (2004) research found that not only do farmworkers endure dangerous and arduous working conditions and the substandard housing mentioned above, but also lack employment security and experience low pay. All of these dynamics result in a complex web within which farmworkers have been historically, and continue to, be marginalized. These evidence claims comprise just an overview of the environmental and structural dimensions of farm labor, and each one requires in-depth research and study, while divorcing any from the others fails to paint a full picture of the dire conditions faced by farmworkers each day and over the course of their work in this sector. Each of these issues manifests itself in different ways across the scales of agriculture, and in what follows; I hope to paint a fuller picture of the angles unique to both conventional and organic agriculture.

Conventional agricultural workers

All of the issues highlighted above are pertinent to those who work in conventional agriculture, whether the farm is small or large. With conventional agriculture, pesticide exposure and pesticide-related illness and death are a huge threat, with 1–5 million pesticide poisonings happening each year across the world, around 20,000 of which are fatal. Also important to note is that these numbers, as touched on previously, are misleadingly low because they do not account for late onset symptoms and illness related to pesticides nor the numerous cases where pesticide-related illness symptoms mask as other illnesses, and therefore go unreported (Harrison, 2011). This issue is further complicated because, as Harrison claims, "poverty, legal status issues, language barriers, political disenfranchisement, and other forms of social marginalization [that are] widespread in farmworking communities tend to obscure pesticide exposures and other problems" (ibid., pp. 9–10). However, as Harrison points out, there is little incentive to increase regulations on pesticide use on the part of farm owners, because reducing the amount of pesticides used increases the amount of labor needed to tend the crops, in turn, increasing the overall cost of labor. From the 1960s–1980s, the United Farm Workers was central to developing early worker protections from agricultural pesticide exposure, but, as detailed in the previous section, began to lose ground (which has never been fully regained) during the Reagan administration in the 1980s (ibid.).

Harrison points out that the shift toward food reform and alternative farming has largely ignored farmworker issues While the focus in the organic movement has centered on personal health and sometimes even community access and well-being, it is still relatively silent on the role of and injustices faced by farmworkers. Many scholars even assert that the organic movement reinforces structural injustice when it comes to farmworkers, contrary to the popular and well-adopted belief that it is somehow inherently more just in all dimensions.

Organic agricultural workers

Labor on organic farms and the issues workers elsewhere in agriculture face intersect. Most of us (myself included before I started studying this topic more in depth), if asked, would certainly make the assumption that laborers in organic agriculture are better off than those in conventional agriculture. However, many scholars argue that justice-related issues proliferate in organic agriculture unseen: people are blinded by what they have come to believe that organic agriculture represents. Further, as I will argue here and later in this book, due to the larger political economic system within which organic agriculture exists, it is difficult for farmers to make ends meet, which has a clear impact on the livelihood of not only their farms and families, but their workers as well. As Margaret Gray poignantly asserts, "Given the constant reinforcement of these benefits, it is easy to see how consumers conflate local, alternative, sustainable, and fair as a compendium of virtues against the demonized factory farm" (2014, p. 41).

First, it is important to better understand and unpack what organic agriculture is. Two predominant definitions shape this sector of agriculture: (1) that provided by the United States Department of Agriculture (USDA); and (2) that of the International Foundation for Organic Agriculture Movements (IFOAM). The USDA definition states:

> Organic agriculture produces products using methods that preserve the environment and avoid most synthetic materials, such as pesticides and antibiotics. USDA organic standards describe how farmers grow crops and raise livestock and which materials they may use. Organic farmers, ranchers, and food processors follow a defined set of standards to produce organic food and fiber. Congress described general organic principles in the Organic Foods Production Act, and the USDA defines specific organic standards. These standards cover the product from farm to table, including soil and water quality, pest control, livestock practices, and rules for food additives.
>
> (USDA, n.d.a)

While the USDA standards cover a variety of issues, they leave out completely any mention of workers or any protections or treatments that they should receive. Many critics of organic agriculture, and specifically USDA organic certification, feel that the nationally adopted definition and standards fall short of providing a context for justice. Obach (2015) points out that enforceable requirements around wages or treatment of workers are absent from the rules of organic certification.

IFOAM's definition was adopted in 2005 and states:

> Organic Agriculture is a production system that sustains the health of soils, ecosystems and people. It relies on ecological processes, biodiversity and cycles adapted to local conditions, rather than the use of inputs with adverse effects. Organic Agriculture combines tradition, innovation and science to benefit the shared environment and promote fair relationships and a good quality of life for all involved.
>
> (IFOAM, n.d.)

One of their key principles, the Principle of Fairness, is characterized by, "equity, respect, justice and stewardship of the shared world, both among people and in their relations to other living beings," and emphasizes that "those involved in Organic Agriculture should conduct human relationships in a manner that ensures fairness at all levels and to all parties – farmers, workers, processors, distributors, traders and consumers" (ibid.). Obach (2015) lauds IFOAM's commitment to social justice as it pertains to workers through their guidelines on access to healthcare, water, fair wages, and the right to organize, as well as a commitment to the prevention of forced labor. However, he acknowledges the lack of support for these efforts due to costly enforcement or accountability, stating, "worker justice remains a vexing issue for the organic movement" (ibid., p. 179).

As Brian Obach further reports, the term organic represents different things across the sector of agriculture, and the USDA and IFOAM definitions just scratch the surface of the variety of ways organic agriculture is framed, viewed, and enacted (not to mention the variability of how it is understood by consumers). He described the roots of the organic movement as a genuine social movement, one whose members believed that the agrarian ideal of family and land would lead to nourishment for communities, both physically and metaphorically. Organic farming evolved from a collection of gardening enthusiasts grounded in an agrarian culture that inspired activism committed to the eradication of synthetic chemical use and environmental degradation due to agricultural practices. This group emerged around the same time as the United Farm Workers (in the 1960s and the 1970s) with their main concerns of fair compensation for farmworkers and health and safety, particularly as it pertained to exposure to toxic pesticides. In this time period there was a synergy among these movements, and the centers of activity as detailed by Obach, became communes and farms, food co-ops, and private alternative restaurants and food stores (ibid., p. 42).

At its roots, organic agriculture is tied to what Harrison (2011) refers to as the agrarian ideal. Central to this ideal is the belief that small-scale organic farms are family farms that do not rely on family labor, which makes the gendered division of labor, racial and ethnic discrimination, and violence that also happen in organic agriculture invisible. She draws on the work of Getz, Brown, and Shreck (2008), claiming that, "This misperception exists about the organic industry because the majority of organic farms are small, and because small-scale farmers circulate widely in organic farming political circles" (Harrison, 2011, p. 157). Therefore, from her perspective, those in this sector of agriculture perpetuate this myth of organic agriculture being ethically sound, even though she states that organic growers are just as likely to oppose basic worker protections as farm owners in conventional agriculture. This reinforces, as Allen (2004) points out, that farmers and consumers are the focus of agrifood movements, rather than improving the conditions of those in less privileged economic positions, in this case, farmworkers. In an interview with Kathleen Maclay for *Berkeley News*, Seth Holmes asserts, "We need solutions for unhealthy living and working conditions that are sustainable for farmworkers and for farmers, who may feel trapped by a globalizing market that does not value local food or fair labor practices" (Maclay, 2014).

Julie Guthman (2004) refers to the agrarian ideal as the agrarian dream, and acknowledges that it is in these seemingly disparate and complex viewpoints that exists the paradox of organic farming. She explores the notion that the federal (USDA) definition of organic and its subsequent certification and standards guidelines lessen or eliminate the chance for anything radical to happen in organic agriculture because they equate, in public political space, organic with the absence of noxious inputs rather than the consideration of ethical processes by which food is grown and harvested. She also expresses the view that as such, organic agriculture perpetuates the very systems it purports to fight against, and has become an extension of capitalist agriculture rather than a true counter-movement. Guthman maintains that farm labor concerns have been pushed aside for the sake of developing organic rules, and this is made more problematic by the dynamics of organic farm labor that differ from those in conventional agriculture.

On the plus side, laborers in organic agriculture are exposed to less pesticides or toxicity and have more opportunities for year-round employment. However, this is due to the more manual nature of harvesting organic food, which requires more grower attention in the field. This data is backed up by multiple reports on organic production, in which farmers identify high labor costs as one of the major obstacles they face (Florida Certified Organic Growers & Consumers, Inc., 2011; Strochlic & Sierra, 2011; Walz, 2004). These reports also emphasize the difficulty on organic farms of recruiting and retaining workers due to low wages, rural locations, lack of benefits, and seasonality of the work. Guthman further reports from her study that the use of labor contractors is still common, that most growers do not provide full-time, year-round employment, wages are low, and there are very few employee-owned collectives or unions within organic agriculture. So, while on the face of it, organic agriculture is understood to be more just overall (and people make the intellectual leap that extends to farmworkers, as Guthman states, "Some consumers choose organic food with the assumption that it is somehow better for farmworkers" (2004, p. 119), it simply replicates many of the same injustices already detailed in this chapter.

Little is formally known about the unique challenges facing farmworkers on organic farms. According to Arcury et al. (2009), three predominant challenges are associated with understanding and adequately addressing the justice issues facing farmworkers. First, limited data exists documenting the current status of health and safety for farmworkers. Second, the data that is available indicates problems for farmworkers and their families' health and safety particularly in the areas of housing, adequate insurance coverage, and protection and training. Third, "agricultural labor policy supports the exploitation of farmworkers, increases the risk of injury and illness, and denies justice" (ibid., p. 223). However, even with these challenges, the authors describe positive trends toward farmworker justice, namely seen in the effects of the work of advocacy organizations, farm labor organizations, and the expansion of community-based research about farmworkers.

Labor on organic farms intersects with the ongoing dialogue about agricultural labor more generally. In his Denmark-based study, Hansen (2004) stated that one aim of organic agriculture is to provide a high standard of living for all involved. He conducted farm visits with nine organic farmers in 2003 to examine significant issues in the work environment. In this study, several themes emerged, including physical strain on farmers and workers, and the administration of organic regulations.

> The organic farmers emphasize, that they themselves have chosen to produce under the organic regulations, and that they value the rules and the control of the rules, because they think it is crucial to their creditability and argumentation in relation to the consumers and the surrounding society. But in practice they find the system incredibly complex and inflexible to maneuver in.
>
> (ibid., p. 2)

Other issues included concern with marketing possibilities, working hours, workload, and social relations with employees, more management demanded of the farmer, the professional pride and pleasures of working as an organic farmer, and the need for a better dialogue and strengthening of collaboration among farmers.

Darnhofer (2005) examined organic farming and development in rural Austria and its shift away from the rural locations and the farms as larger, more powerful retailers have become involved in the market. In that scenario, small-scale agriculture ultimately loses out. However, he also argues that this need not necessarily be the case. Diversification can help a farmer's ability to compete in local markets and expand income sources. However, entering into the organic market's regulations and rules can limit organic farmers from gaining economically from participating in that market unless done on a scale big enough to benefit. The reason for this is that marketing challenges result in organic products seen oftentimes as conventional so they can be sold alongside those products, increasing the amount of produce or goods available from which to profit. Finally, Darnhofer raises the important question of whether organic farmers who market their products through corporate retailers do contribute to rural development.

Jansen (2000) conducted an informative literature review highlighting the differences between conventional and organic agriculture in regard to labor, and how conversion from conventional to organic affects workers and farmers alike. He discusses labor use changes throughout Western Europe during the conversion to organic to make the argument that more labor is needed (leading to higher rural employment) but that high labor costs keep the organic sector from further development. Also, Jansen points out that as the organic supply increases, the premium that consumers pay will decrease, creating further pressure on the organic sector. He finds a diversity of "livelihood and work strategies" (ibid., p. 249) that are organized around social relations, such as friendship and households, as defining organic agricultural production in Europe.

Jansen cites Whatmore (1991) who distinguished four labor circuits: (1) agricultural labor (manual or administrative) that is directly related to the cultivation of plants and animals; (2) domestic household labor; (3) non-agriculture farm labor (running farm shops, educational programs, camp sites); and (4) off-farm wage labor consisting of off-farm paid employment that generates income for the budget. All of these are intertwined and not mutually exclusive. He further finds that quality of labor is dependent upon four key factors: (1) the content of work (possibilities of defining tasks, acquiring knowledge); (2) labor relations (such as gender differences); (3) working conditions (health and safety, intensity of workload); and (4) the terms of employment (pay, insurance, benefits, etc.).

In addition, Jansen points out how comparisons between organic and conventional farms are difficult for several reasons (How should they be grouped? According to size? Capital? Crops?), the type of farm activity (most times the high-labor, high-wage crops are being converted to organic), the time of measurement (startup labor may increase after conversion but it goes back down over time), and farm size (costs of labor are much higher on small farms after conversion). His study also examined labor satisfaction by asking whether shifting from conventional to organic has been a shift in "lifestyle ideology or material experience" (Jansen, 2000, p. 263). In response, 63.9 percent of farm managers indicated greater satisfaction after converting to organic. He reinforces the evolving landscape of ideology behind conversion by offering that in the 1970s many people became farmers because they wanted to demonstrate through their lives and work how they thought the economy and society should be structured. Overall, the "risk in organic farming results from the prohibition of simple applications of biocides, the use of many techniques which are still experimental, lower yields, uncertain markets, uncertain price premiums, and difficulties with social acceptability in the local community" (ibid., p. 264). Now it is less about the ideology that brought people to farming in the 1970s and more about rural people redefining the practice of agriculture.

Because of this trend, organic farm managers and workers have become more educated than those on conventional farms.

In a study by Cross, Edwards, Hounsome, and Edwards-Jones (2008), researchers conducted a comparison of self-reported health of workers on organic versus conventional farms in the United Kingdom (90+ percent were migrant workers from other countries). They found no significant difference for three of the instruments they used to measure worker health (Short Form 36, EuroQol EQ-5D and the Visual Analogue Scale), whereas organic farmworkers scored higher on a fourth health instrument called the Short Depression Happiness Scale (SDHS). This indicates that workers on organic farms were happier than their counterparts working on conventional farms, which correlated to organic workers having a wider range of tasks.

In a survey of 1144 organic farms in the United Kingdom and Ireland, Morison, Hine, and Pretty (2005) set out to assess whether organic agriculture provides more opportunity for labor. They found that organic farms employ 135 percent more full-time equivalent (FTE) jobs per farm than conventional farms. However, the mean number of workers per area was lower for organic farms, due to size. They conclude that if 20 percent of all farms in both countries were organic, the number of available farming jobs in each would increase.

In a pivotal US-based study on labor, equity, and justice in organic agriculture, Amy Shreck, Christy Getz, and Gail Feenstra (2006) found little consensus among farmers about the idea of including justice for workers as part of the certification process for farms. They used a survey and in-depth interviews about the incorporation of social standards into organic certification criteria, highlighting that the question remains whether organic farms provide better working conditions than do conventional farms. In the anonymous mail questionnaire sent to a random sample of 500 organic farmers in California, 188 farmers returned the survey (39 percent). The sample was selected from 1762 organic farmers provided by the California Department of Food and Agriculture, and includes only organic farms plus those who combine organic with conventional techniques. The survey focused on farmers' thoughts on social sustainability in organic agriculture, including farm activities, hiring and employment practices, beliefs regarding sustainability, and demographic characteristics.

In the Shreck et al. study, small-scale farms (50 acres or less, $50,000 or less in annual sales) made up three-fourths of the respondents. The respondents indicated relatively little support for adding social certification standards to the certification requirements. Only 42.3 percent agreed or strongly agreed that organic certification should include criteria on working conditions, and interestingly those who did agree/strongly agree tended to be farmers who do not employ farmworkers (i.e. those who do employ them were the ones who believed that it should not be included). They also found that large farms pay more and offer more benefits. They elaborate, stating that,

> As a number of farmers reiterated through comments they included on their questionnaires, even if they believe that organic agriculture should

> ensure fair and healthy working conditions for farmworkers, they explain that it is simply not economically viable given the realities of the market.
>
> (ibid., pp. 444–445)

Many more expressed the concern that the financial burden of requirements for labor makes them untenable. Some farmers argued that higher labor standards should be required for both conventional and organic farms, not simply organic. Many farmers commented on how they cannot provide something for workers that they cannot even have for themselves, such as benefits and higher pay.

Shreck et al. (2006) discuss three phenomena that help in explaining why it is unclear that organic agriculture is more socially sustainable for farmworkers than conventional: (1) the social dimension of sustainability is widely interpreted; (2) the costs of making organic agriculture socially sustainable are being externalized; and (3) there are individuals whose practices are atypical but do demonstrate that organic production can be environmentally, economically, and socially sustainable. They argue that we need to further examine positive examples to show farmers a path to being more in alignment with these three principles/vectors of sustainability.

In their investigation of the intersection of farm labor and organic certification and assessment of organic farmers' attitudes regarding the social dimension of sustainability, Shreck et al. find that while organic farmers broadly accepted the idea of social justice associated with their practice, they did not predominantly support the practices that reinforce justice, such as the right to organize. Furthermore:

> The full costs of making organic agriculture socially sustainable are being externalized to a certain degree. This is suggested by the reality that most employers of farmworkers do not provide, and perceive that they cannot afford to provide, things like living wages and health insurance.
>
> (ibid., p. 446)

All of these findings have important implications, and contribute to the complexity inherent in labor as it relates to organic agriculture, about which so many hold an ideal of justice and fairness. In their report, Hoodes et al. state:

> Concerns about farmer and farmworker rights, migrant labor, changing organic contracts, and farmer and worker wages and benefits are now being challenged. The need to find ways to institutionalize fair prices, wages and benefits; to build bridges with the worker community, and to address scale, ownership, and control of the organic sector are all viewed as critical to the long-term success and sustainability of organic agriculture.
>
> (2010, p. 28)

As they go on to state, all of this is tied up in the market, and the assurance of fair market prices that make their way back to the farmers and farmworkers on these organic farms.

Conclusion

In 1992, Frank L. Farmer and Doris P. Slesinger carried out a literature review, "Health Status and Needs of Migrant Farm Workers in the United States," and it is fair to say that what they uncovered in their piece applies not only to migrant workers, but also to workers across the agricultural system, both in scale, and in both conventional and organic. In addition, not much has changed in the nearly 25 years since their report was released. They highlight issues surrounding acute illness, chronic illness, lack of preventive care, occupational and environmental hazards, the barriers to care farmworkers face (and lack of understanding of many healthcare providers). In addition, they state, "Farmworkers lead a hard life filled with strenuous work, stress, and anxiety about employment; live under substandard conditions, and rarely get the health care they require" (Farmer & Slesinger, 1992, p. 233).

While the issues detailed in this chapter paint a dire picture of life as an American farmworker, it is important to note that the context within which these conditions were created has developed and perpetuated itself over time. Each one of the issues detailed here has its roots in a political economic landscape that has been fertile for oppression and marginalization of farmworkers. While we have just explored the process claims inherent in the social dimensions of who works in agriculture in the United States, and the evidence claims or environmental and structural dimensions of the issues that face them across different scales of agriculture, in Chapter 3 we will explore more fully the process claim questions of "why?" and "how did this come to be?" through the presentation of a brief history of and the impact of policy on farmworkers in the United States.

Notes

1 According to www.agcensus.usda.gov/Publications/Organic_Survey/:

> The 2014 Organic Survey results reflect the industry as of the time the mail list was created and for the 2014 production year. Data users should allow for differences when comparing data between the 2014 survey and the 2011 Certified Organic Production Survey or the 2008 Organic Production Survey. The 2011 survey collected data only from certified operations, whereas the 2014 survey collected data from certified, exempt, and transitioning operations. The 2008 survey collected information from the same organic groups (certified, exempt, and transitioning) as the 2014 Organic Survey, but the 2014 survey was more comprehensive."

2 According to www.agcensus.usda.gov/Publications/Organic_Survey/, it is important to note that

> The 2014 Organic Survey results reflect the industry as of the time the list was built and the 2014 production year. Comparisons with the 2012 Census of Agriculture, other NASS publications and other non-NASS sources must allow for differences in reference periods, organic definitions and weighting methodologies.

Therefore, these percentages are calculated and used here just to give a general sense of how many farms/acres are utilized for organic farming.

References

Allen, P. (2004). *Together at the table: sustainability and sustenance in the American agrifood system*. University Park, PA: The Pennsylvania State University Press.

Arcury, T. A., Quandt, S. A., & Russell, G. B. (2002). Pesticide Safety among Farmworkers: Perceived Risk and Perceived Control as Factors Reflecting Environmental Justice. *Environmental Health Perspectives Supplements*, 110, 233–240. Available at: http://doi.org/Article

Arcury, T. A., Wiggins, M. F., & Quandt, S. A. (2009). Conclusions: An Agenda for Farmworker Social Justice in the Eastern United States. In S. A. Quandt, & T. A. Arcury (Eds.), *Latino farmworkers in the Eastern United States* (pp. 221–233). New York: Springer. Available at: www.springerlink.com/content/k4138470 31p03243/abstract/

Barndt, D. (2002). *Tangled routes: women, work, and globalization on the tomato trail*. Lanham, MD: Rowman & Littlefield.

Bon Appétit Management Company Foundation and United Farm Workers. (2011). Farmworker Inventory: Bon Appétit Management Company (Inventory No. 1). Available at: www.bamco.com/uploads/documents/farmworkerinventory_0428_2011.pdf

Brown, S., & Getz, C. (2011). Farmworker Food Security and Production of Hunger in California. In A. H. Alkon, & J. Agyeman (Eds.), *Cultivating food justice: race, class, and sustainability* (pp. 121–146). Cambridge, MA: MIT Press.

Cancino, A. (2016). Uphill Battle to Get Seasonal Farmworkers Health Insurance. *The San Diego Union-Tribune*. Available at: www.sandiegouniontribune.com/news/2016/jun/18/uphill-battle-to-get-seasonal-farmworkers-health/ (accessed July 26, 2016).

Carroll, D., Georges, A., & Saltz, R. (2011, May). Changing Characteristics of U.S. Farm Workers: 21 Years of Findings from the National Agricultural Workers Survey. Paper presented at the Implications for Farmers, Farm Workers, and Communities: Immigration Reform and Agriculture Conference, University of California, D.C. Campus. Available at: https://migrationfiles.ucdavis.edu/uploads/cf/files/2011-may/carroll-changing-characteristics.pdf

Carroll, D., Samardick, R., Bernard, S., Gabbard, S., & Hernandez, T. (2005). *Findings from the National Agricultural Workers Survey (NAWS) 2001–2002: A demographic profile of United States farm workers* (No. 9) (pp. ix–67). Washington, DC: U.S. Department of Labor.

CDC-NIOSH. (2005). *Pesticide-Related Illness and Injury Surveillance: A How-To Guide for State-Based Programs (2006–102)*. Available at: www.cdc.gov/niosh/docs/2006-102/ (accessed January 21, 2012).

Cross, P., Edwards, R. T., Hounsome, B., & Edwards-Jones, G. (2008). Comparative Assessment of Migrant Farm Worker Health in Conventional and Organic Horticultural Systems in the United Kingdom. *The Science of the Total Environment*, 391(1), 55–65.

Darnhofer, I. (2005). Organic Farming and Rural Development: Some Evidence from Austria. *Sociologia Ruralis*, 45(4), 308–323.

Farmer, F. L., & Slesinger, D. P. (1992). Health Status and Needs of Migrant Farm Workers in the United States: A Literature Review. *Journal of Rural Health*, 8(3), 227–234. Available at: http://doi.org/10.1111/j.1748-0361.1992.tb00356.x

Farmworker Justice. (2013). Exposed and Ignored: How Pesticides Are Endangering Our Nation's Farmworkers. Available at: www.farmworkerjustice.org/sites/default/

files/aExposed%20and%20Ignored%20by%20Farmworker%20Justice%20singles%20compressed.pdf

Farmworker Justice & The National Center for Farmworker Health. (2015). Farmworkers' Health Fact Sheet: Data from the National Agricultural Workers Survey. Available at: www.ncfh.org/uploads/3/8/6/8/38685499/fs-nawshealthfactsheet.pdf

Flocks, J., Kelley, M., Economos, J., & McCauley, L. (2011). Female Farmworkers' Perceptions of Pesticide Exposure and Pregnancy Health. *Journal of Immigrant and Minority Health*, 14(4), 626–632. Available at: http://doi.org/10.1007/s10903-011-9554-6

Florida Certified Organic Growers & Consumers, Inc. (2011). *Organic perspectives: understanding the views of Florida consumers, specialty crop farmers, and retailers*. Gainesville, FL: Florida Certified Organic Growers & Consumers, Inc.

Galarneau, C. (2013). Farm Labor, Reproductive Justice: Migrant Women Farmworkers in the US. *Health and Human Rights*, 15(1), 144–160.

Getz, C., Brown, S., & Shreck, A. (2008). Class Politics and Agricultural Exceptionalism in California's Organic Agriculture Movement. *Politics & Society*, 36(4), 478–507.

Gorton, J., & van Hightower, N. R. (1999). Intimate Victimization of Latina Farm Workers: A Research Summary. *Hispanic Journal of Behavioral Sciences*, 21(4), 502–507. Available at: http://doi.org/10.1177/0739986399214008

Gray, M. (2014). Labor and the Locavore: The Making of a Comprehensive Food Ethic. Available at: http://librarytitles.ebrary.com/Doc?id=10779332

Griffith, D. (2006). *American guestworkers: Jamaicans and Mexicans in the U.S. labor market*. University Park, PA: Pennsylvania State University Press. Available at: http://site.ebrary.com/id/10532215

Guthman, J. (2004). *Agrarian dreams: the paradox of organic farming in California*. Berkeley, CA: University of California Press. Available at: http://site.ebrary.com/id/10064761

Hansen, L. (2004, March). What Is the Working Environment Like on Organic Farms? *Forskningsnytt om økologisk landbruk i Norden*. Available at: http://orgprints.org/2363/ (accessed September 22, 2013).

Harrison, J. L. (2011). *Pesticide drift and the pursuit of environmental justice*. Cambridge, MA: MIT Press.

Holden, C. (2002). Bitter Harvest: Housing Conditions of Migrant and Seasonal Farmworkers. In C. D. Thompson, & M. F. Wiggins (Eds.), *The human cost of food: farmworkers' lives, labor, and advocacy* (pp. 169–193). Austin, TX: University of Texas Press.

Holmes, S. M. (2013). *Fresh fruit, broken bodies: migrant farmworkers in the United States*. Berkeley, CA: University of California Press.

Hoodes, L., Sligh, M., Behar, H.*et al.* (2010). *National Organic Action Plan: from the margins to the mainstream – advancing organic agriculture in the U.S.* Pittsboro, NC: Rural Advancement Foundation International-USA.

IFOAM (International Federation of Organic Agriculture Movements) (n.d.) Leading, Uniting and Assisting the Organic Movement since 1972. Available at: www.ifoam.org

Jansen, K. (2000). Labour, Livelihoods and the Quality of Life in Organic Agriculture in Europe. *Biological Agriculture & Horticulture*, 17(3), 247–278. Available at: http://doi.org/10.1080/01448765.2000.9754845

Kandel, W. (2008). USDA Economic Research Service – ERR60. Available at: www.ers.usda.gov/publications/err-economic-research-report/err60.aspx (accessed January 31, 2016).

Kelley, M. A., Flocks, J. D., Economos, J., & McCauley, L. A. (2013). Female Farmworkers' Health during Pregnancy: Health Care Providers' Perspectives. *Workplace Health & Safety*, 61(7), 308–313. Available at: http://doi.org/10.1177/216507991306100706

Kiehne, E., & Mendoza, N. S. (2015). Migrant and Seasonal Farmworker Food Insecurity: Prevalence, Impact, Risk Factors, and Coping Strategies. *Social Work in Public Health*, 30(5), 397–409. Available at: http://doi.org/10.1080/19371918.2015.1019173

Maclay, K. (2014, March 6). Berkeley Researcher Examined, Lived the Life of a Migrant Farmworker. *Berkeley News*. Available at: http://news.berkeley.edu/2014/03/06/berkeley-researcher-lived-the-life-of-a-migrant-farmworker/ (accessed February 2, 2016).

Marquez, E. C., & Schafer, K. S. (2016). *Kids on the frontline: How pesticides are ruining the health of rural children*. Pesticide Action Network North America. Available at: www.panna.org/sites/default/files/KOF-report-final.pdf

May, J. J. (2009). Occupational Injury and Illness in Farmworkers in the Eastern United States. In T. A. Arcury, & S. A. Quandt (Eds.), *Latino farmworkers in the Eastern United States* (pp. 71–101). New York: Springer.

Morison, J., Hine, R., & Pretty, J. (2005). Survey and Analysis of Labour on Organic Farms in the UK and Republic of Ireland. *International Journal of Agricultural Sustainability, 3*(1), 24–43. Available at: http://doi.org/10.1080/14735903.2005.9684742

National Center for Farmworker Health. (2012). Farmworker Health Factsheet: Facts about Farmworkers. Available at: www.ncfh.org/uploads/3/8/6/8/38685499/fs-facts_about_farmworkers.pdf

National Farm Worker Ministry. (n.d.a). Low Wages. Available at: http://nfwm.org/education-center/farm-worker-issues/low-wages/ (accessed January 26, 2016).

National Farm Worker Ministry. (n.d.b). Women's Issues. Available at: http://nfwm.org/education-center/farm-worker-issues/womens-issues/ (accessed February 14, 2016).

Obach, B. K. (2015). *Organic struggle: the movement for sustainable agriculture in the United States*. Cambridge, MA: The MIT Press.

Reeves, M., Katten, A., & Guzman, M. (2002). *Fields of poison* (pp. 1–38). California: Pesticide Action Network, California Rural Legal Assistance Foundation, and United Farm Workers of America. Available at: www.panna.org/sites/default/files/FieldsofPoison2002Eng.pdf

Rothenberg, D. (1998). *With these hands: the hidden world of migrant farmworkers today*. New York: Harcourt Brace & Co.

Runkle, J., Culley, J., Economos, J., *et al.* (2014). Participatory Educational Intervention to Address Heat, Ergonomic Stress, and Pesticide Exposure as Workplace Hazards for Female Farmworkers in Central Florida. *Journal of Agromedicine*, 19(2), 235–265. Available at: http://doi.org/10.1080/1059924X.2014.892448

Runkle, J., Tovar-Aguilar, J. A., Economos, E., *et al.* (2013). Pesticide Risk Perception and Biomarkers of Exposure in Florida Female Farmworkers. *Journal of Occupational and Environmental Medicine*, 55(11), 1286–1292.

Rust, G. S. (1990). Health Status of Migrant Farmworkers: A Literature Review and Commentary. *American Journal of Public Health*, 80(10), 1213–1217. Available at: http://doi.org/10.2105/AJPH.80.10.1213

Seif, H. (2008). Wearing Union T-Shirts: Undocumented Women Farm Workers and Gendered Circuits of Political Power. *Latin American Perspectives*, 35(1), 78–98. Available at: http://doi.org/10.1177/0094582X07310969

Shreck, A., Getz, C., & Feenstra, G. (2006). Social Sustainability, Farm Labor, and Organic Agriculture: Findings from an Exploratory Analysis. *Agriculture and Human Values*, 23(4), 439–449.

Sologaistoa, E. (2011). *Farmworkers in the Southeast: Alabama, Florida, Georgia, Mississippi* (pp. 1–30). Tallahassee, FL: Florida Association of Community Health Centers.

Southern Poverty Law Center. (n.d.). *Sexual violence against farmworkers: a guidebook for criminal justice professionals*. Available at: www.splcenter.org/20100401/sexual-violence-against-farmworkers-guidebook-criminal-justice-professionals (accessed February 14, 2016).

Strochlic, R., & Sierra, L. (2011). *Conventional, mixed and "deregistered" organic farmers: entry barriers and reasons for exiting organic production in California*. Davis, CA: California Institute for Rural Studies. Available at: www.cirsinc.org/publications/farm-labor?download=7:conventional-mixed-and-qderegisteredq-organic-farmers-entry-barriers-and-reasons-for-exiting-organic-production-in-california-

Thompson, C. D. (2002). Introduction. In *The human cost of food: farmworkers' lives, labor, and advocacy* (pp. 2–19). Austin, TX: University of Texas Press.

UFW (United Farm Workers of America). (n.d.). Home page. Available at: www.ufw.org/_page.php?menu=research&inc=_page.php?menu=research&inc=history/01.html (accessed January 26, 2016).

USDA (U.S. Department of Agriculture), National Agricultural Statistics Service. (2008). *2007 census of agriculture: 2008 organic production survey (No. 1)*. Available at: www.agcensus.usda.gov

USDA (U.S. Department of Agriculture), National Agricultural Statistics Service. (2012). *2011 certified organic production survey*. Available at: www.agcensus.usda.gov/Publications/Organic_Production_Survey/

USDA (U.S. Department of Agriculture), National Agricultural Statistics Service. (2014). *2012 census of agriculture: United States summary and state data* (Vol. 1, Part 51). Available at: www.agcensus.usda.gov/Publications/2012/Full_Report/Volume_1,_Chapter_1_US/usv1.pdf

USDA (U.S. Department of Agriculture), National Agricultural Statistics Service. (2015). *Organic production survey (2014)* (Special Studies Vol. 3, Part 4). Available at: www.agcensus.usda.gov/Publications/2012/Online_Resources/Organics/ORGANICS.pdf

USDA (U.S. Department of Agriculture). *Certified USDA organic agriculture* (n.d.a). Available at: www.solgardens.net/certified-usda-organic.html (accessed October 10, 2016).

USDA (U.S. Department of Agriculture), Economic Research Service. (n.d.b). *ARMS farm financial and crop production practices*. Available at: www.ers.usda.gov/data-products/arms-farm-financial-and-crop-production-practices.aspx#.U8agjvldUbs (accessed January 3, 2016).

USDA (U.S. Department of Agriculture), Economic Research Service. (n.d.c). *Farm household well-being*. Available at: www.ers.usda.gov/topics/farm-economy/farm-household-well-being.aspx (accessed January 3, 2016).

USDA (U.S. Department of Agriculture), Economic Research Service. (n.d.d). *Farm labor: background*. Available at: www.ers.usda.gov/topics/farm-economy/farm-labor/background.aspx (accessed December 30, 2015).

Walz, E. (2004). *Final results of the fourth national organic survey: sustaining organic farms in a changing organic marketplace*. Santa Cruz, CA: Organic Farming Research Foundation.

Whatmore, S. (1991). *Farming women: gender, work, and family enterprise*. Basingstoke: Macmillan.

3 A brief history and impact of policy on laborers in U.S. agriculture

Process claims: the social dimensions of farm labor

As illustrated in Chapter 2, farmworkers in the United States live and work in often dire circumstances, and are a population that is largely invisible, even in food movements. Further, one does not have to dig very deeply to find that little has changed regarding the predicaments that farmworkers face: published accounts from the 1940s and 1950s look similar in many ways to those today. In other ways, their situations are worse, as farmworkers today are not just comparatively voiceless due to their status as workers but are burdened by added layers of voicelessness as free trade and guestworker policies have increased the number of immigrants (documented and those undocumented) working in agriculture. While, as Chapter 1 asserts, the environmental justice movement intersects with both broader environmental and labor movements, this crossroads has not been leveraged effectively to move the needle toward substantial change for farmworkers. As Brian Obach states:

> Labor unions and environmental movement organizations are among the most powerful social movement sectors in the US when they act together, they can advance policies that protect both working people and the natural environment. Yet divisions between these two actors can yield environmental devastation and attacks on the interests of workers and their unions.
>
> (2004: 7)

In addition, labor unions have been hit hard in recent years throughout various levels of the political landscape.

When seeking to explore strategies for change that may be effective across multiple scales of agriculture, it is important to understand not only what issues farmworkers face today, but why they are facing these issues. In the context of agricultural labor, policy has had direct and indirect influences. Direct influences include the exclusion of farmworkers from national labor protection policies, the lack of policy on and enforcement of workplace and housing safety regulations, and legislation surrounding immigrant labor streams in agriculture. When considering why farmworkers are excluded from legislation that protects or provides benefits for other classes of workers, Daniel Rothenberg states:

> There is some mention of the political obstacles to protecting agricultural workers, a few comments about the government's intention to include farmworkers in the future, and the occasional remark that excluding farmworkers from basic worker protections is simply unfair and unjust. Ultimately, the decision [to exclude them] reflects the gross imbalance between the political influence of agricultural interests and the powerlessness of farmworkers.
>
> (1998, p. 214)

Wiggins (2009) uses the phrase "agricultural exceptionalism" to describe the phenomenon of excluding farmworkers from legal protections that benefit other workers. Sarah Rodman and her colleagues (Rodman et al., 2016, p. 2) highlight that the "legal protections concerning minimum wage, overtime pay, unemployment insurance, collective organizing and bargaining, and occupational health all contain exceptions for farmworkers." This chapter will showcase how this exceptionalism is seen in a number of different policy arenas.

Less direct, but important and pivotal influencing factors include free trade policies and their impacts. These include the U.S. farmer/farm corporations' ability to import cheap labor, from countries included in these agreements, and the flood of cheap goods and services to their markets, which in turn affects the possibility of earning a viable livelihood in those countries. As Sandy Brown and Christy Getz (2011) note, trade policy not only manages labor flow, but also creates an exploitable 'other,' someone who has few options but to comply.

In their concluding chapter for *Latino Farmworkers in the Eastern United States: Health, Safety, and Justice*, Arcury, Wiggins, and Quandt (2009) lay out the key areas for policy change when it comes to moving the needle toward justice for farmworkers: labor policy, housing policy, pesticide policy, healthcare policy, and immigration policy. Not surprisingly, these areas directly overlap with those that exclude farmworkers completely or fail to consider the implications for the farmworker population. Attention is needed to create or bolster policy in these areas, as are better enforcement and regulation even where formal policy is comparatively strong. As such, considerations of enforcement are just as crucial as the policies themselves in ensuring justice for workers.

How did this come to be? A brief history of policy and U.S. farmworkers

This section reviews twentieth- and twenty-first-century[1] national policies that continue to impact farmworkers' livelihoods and access to just working and living conditions.[2] I begin with a review of major policies impacting labor relations and standards, along with those regarding health and safety, with the latter related to pesticide use and restriction. Then, I explore the impacts of immigration and trade policy on farmworkers—in its influence on who can work in agriculture in the United States, how they come to work in it, and the limitations of their rights as non-U.S. citizens. Following is a consideration of the impact of policy, certification, and regulation in organic agriculture on laborers in that sector. In examining this, keep in mind the landscape of agricultural labor in the United States as presented in Chapter 2. The convergence of this complex system of policies is key to furthering the process claim question "How did this come to be?" and is one fruitful area for justice claims around how things *should* be, the focus of Chapter 6.

Labor relations and standards

During the Great Depression, Congress passed the Wagner Act of 1935 to govern worker organizing and collective bargaining, and created the National Labor Relations Board (NLRB) as the overseeing body (Wiggins, 2009). After the Act's passage, the NLRB made the decision to draw "a distinction between field laborers and processing laborers. The former group was excluded from the act; the latter, included" (Hahamovitch, 1997, p. 153). This distinction effectively excluded farmworkers from the protections afforded by collective bargaining, and aligned the benefits of the policy with industrial workers. Margaret Gray argues that these decisions were driven largely by farmers and farmer organizations arguing for agricultural exceptionalism with the rationale that "no other industry … faces the same demands or deserves similar treatment, particularly in regard to safeguarding its cheap labor supply" (2014, p. 23). Rothenberg states, "excluding agricultural workers from the NLRA dealt a harsh blow to struggling farmworker unions … and further distanced farmworkers from industrial workers, whose unions grew in size and political influence" (1998, pp. 213–214). Not only would the ability to organize eventually give voice to farmworkers, but it would allow them to take positions on and make progress toward fair wages, job security, and the enforcement of safety and decent sanitation standards. Additionally, this organizing would create a formal grievance procedure through which to report if and when their rights are violated (National Farm Worker Ministry, n.d.). But that did not happen.

In his book, *From the Jaws of Victory: The Triumph and Tragedy of Cesar Chavez and the Farm Worker Movement*, Matthew Garcia writes, "The task of extending these rights to farm workers would fall to a new generation of activists, most famously Cesar Chavez and the many people responsible for building the United Farm Workers Union in the 1960s and early 1970s" (2012, p. 2). However, as Garcia goes on to note, Chavez's stance on the Act was complex, and "he publicly lamented the exclusion of farm workers from collective bargaining rights but privately relished the freedom to use the boycott whenever and wherever he saw fit" (ibid., p. 127). Chavez went on to work with the California state legislature to create the Agricultural Labor Relations Act (ALRA), which afforded the same rights to agricultural workers that were provided to industrial workers by the NLRA while addressing some of his reservations. The ALRA addressed these concerns by allowing for secondary boycotts and regulating elections during the peak of the agricultural season so migrant workers could participate.

The Agricultural Employee Exemption from the Fair Labor Standards Act (FLSA) of 1938 further complicates policy affecting agricultural laborers. Ten classes of employees found themselves left out of the concept of fair labor as birthed in this Act, two of whom were agricultural workers and domestic workers. This Act, part of President Roosevelt's second wave of New Deal policies, governs minimum wage rates, overtime provisions, and child labor laws, and was meant to ensure workers the rights necessary to earn a safe and decent living wage so they could fully participate in the economy (Wiggins, 2009). More specifically, it guaranteed a minimum wage for each hour worked and required overtime pay to employees working more than 40 hours in a week (overtime pay is one and a half times their regular rate of pay for each hour exceeding 40) (Harris, 2005; National Farm Worker Ministry, n.d.). Why were farmworkers and domestic workers left out? As Anderson explains:

> [T]he bill rests squarely upon the interstate commerce clause of the Constitution … businesses of a purely local type which serve a particular local community, and which do not send their products into the streams of interstate commerce, can be better regulated by the laws of the communities and of the States in which the business units operate.
>
> (1988, p. 653)

Therefore, the regulations were left to those beneath the federal level with the argument that farms were not, or at least not in any major way, engaged in interstate commerce. With the decision to exclude agricultural workers from the protections afforded under this Act, Congress ensured that farmworkers lacked the right to fair and equitable labor practices in their field of work. The FLSA was updated in 1966 with provisional inclusions for farmworkers, but, as the National Farm Worker Ministry (n.d.) points out:

> To this day it continues to exclude them in significant ways: farm workers have no right to overtime pay, workers on small farms are not entitled to receive minimum wage, and children as young as twelve are legally allowed to work in the fields.

Moreover, "there are no restrictions on agricultural work being done by children as young as 12 years old early in the morning or late into the night" (Farmworker Justice, n.d.). All of this contributes to, as Gray (2014, p. 49) puts it, "the institutional marginalization of agricultural workers – including the denial of basic labor and human rights – that reproduces their inequality in the workplace and confines them to substandard working conditions."

Additionally, the Social Security Act of 1935 "originally excluded all farmworkers, denying them unemployment, old-age, and disability benefits. In 1950, the law was amended to include many agricultural workers" (Coalition of Immokalee Workers, n.d.). Larry DeWitt (2010), in a paper published on the Social Security Administration website, explores the oft-suspected racial bias rationale for the exclusion of certain classes of workers, insofar that many scholars believe the exclusion of workers such as those in agriculture was motivated by the fact that many people in the excluded sectors were people of color. He concludes that it was not overt racial bias, but that "the exclusion of agricultural and domestic workers from the early program was due to considerations of administrative feasibility involving tax-collection procedures." He goes on to assert that farm owners did not want to cover their employees because they did not want to pay taxes into the system that would ultimately support Social Security, and further states that there is little to no information on the feelings of farmworkers at that time about their inclusion in the Social Security Act. Another consideration was the difficulty of implementation when it came to providing this benefit to farmworkers, and of collecting the requisite taxes due to the movement of workers continually through the system. The 1950 expansion of the Act provided for "coverage of regularly employed farm workers based on 60 full days of work during a quarter if workers had continuous employment with the same employer during a preceding 3-month period" (Cohen & Myers, 1950). This provided an answer to the complex tension of the applicability of Social Security for farmworkers due to the aforementioned dynamics of disinterest and complications of inclusion, but continues to exclude a large part of the agricultural labor force.

This same conundrum of grower and producer buy-in and difficulty of implementation also is central to debates about unemployment insurance. Due to the seasonal nature of farm work in many areas, the payroll threshold at which farm owners have to provide and pay into this type of insurance may influence the wages and optional benefits they are willing or able to provide. While unemployment insurance is mandated under the Federal Unemployment Tax Act (FUTA), oversight is done at the state level. The agricultural sector has special regulations to start this, and many states exempt small farmers from providing UI or exclude non-immigrant temporary H-2A (agricultural) guest-workers from coverage (Bon Appétit Management Company Foundation and United Farm Workers, 2011). Such exemptions affect worker stability, the length of unemployment between jobs, and the wages workers will accept to find new employment. Farmworkers not covered by unemployment insurance tend to spend less time without work than those not covered in other sectors, but upon reemployment are willing to work for lower wages (Kandilov & Kandilov, 2010).

Even implementation of the Civil Rights Act of 1964 differs when it comes to farmworkers. Title VII of the Act:

> Makes it illegal to discriminate against someone on the basis of race, color, religion, national origin, or sex. The law also makes it illegal to retaliate against a person because the person complained about discrimination, filed a charge of discrimination, or participated in an employment discrimination

> investigation or lawsuit. The law also requires that employers reasonably accommodate applicants' and employees' sincerely held religious practices, unless doing so would impose an undue hardship on the operation of the employer's business.
>
> (EEOC, n.d.a)

This Act established the Equal Employment Opportunity Commission (EEOC) (EEOC, n.d.b) to oversee its implementation and violations where they take place. As illustrated in Chapter 2, farmworker women face a high incidence of sexual harassment and violence, with little opportunity for recourse. While the text of the law would suggest positive changes in the working environment for farmworker women, it did little in practice. In her report for Oxfam America, *Working in Fear: Sexual Violence Against Women Farmworkers in the United States*, Sara Kominers points out:

> Although Title VII of the Civil Rights Act of 1964 theoretically protects all workers, including unauthorized workers, from sexual violence, it is often unable to provide women farmworkers with meaningful protections because it fails to address the combination of discrimination factors they face.
>
> (2015, p. 3)

Kominers goes on to state that a combination of factors including acculturated subservience to men based on norms and customs in their home country, lack of citizenship, and fears of deportation, in addition to earning less than their male counterparts, puts women farmworkers in a uniquely dangerous and exploitable position.

Health, safety, and worker protection

Laws on health, safety, and worker protection also are applied in disparate ways when it comes to farmworkers. According to the U.S. Department of Labor, between 2003 and 2011, nearly 6,000 farmworkers died due to work-related injuries, and each day nearly 250 agricultural workers experience an injury that results in lost working time, an injury rate over 40 percent of the rate for all other workers (OSHA. n.d.b). Considering these numbers as reported by the federal government and keeping in mind that workplace injury and safety violations are underreported by agricultural laborers for the reasons explored in Chapter 2, it becomes clear that policy regulation around worker health and safety is central to justice for farmworkers.[3]

The Occupational Health and Safety Act (OSHA), enacted in 1970, did not specifically include farmworker and farm-related safety issues (Wiggins, 2009). This Act was:

> To assure safe and healthful working conditions for working men and women; by authorizing enforcement of the standards developed under the

> Act; by assisting and encouraging the States in their efforts to assure safe and healthful working conditions; by providing for research, information, education, and training in the field of occupational safety and health; and for other purposes.
>
> (OSHA, n.d.a)

While in many ways this Act was positioned to help farmworkers work in safer conditions, it exempted all small businesses, including farms, with fewer than 11 employees from its standards. In 1987, OSHA expanded to include the Farmworker Standard, which included a Field Sanitation Standard requiring the provision of toilets, water, and hand-washing stations in the fields. However, again, small farms, or those with fewer than 11 employees, were exempted (OSHA, n.d.d; Wiggins, 2009). As a result, says Harris, OSHA is inadequate in addressing working conditions on many of the farms nationwide with ten or fewer employees, adding, "Farm work is a dangerous occupation despite the size of the farm and to lack the ability to regulate the workplace allows certain farmworkers to be in danger from exposure to hazardous materials and hazardous conditions" (2005, pp. 16–17). Also, as Gray (2014, p.180) states, "State laws differ from federal ones, including OSHA laws. Employers are required to follow all laws and adhere to the stricter standard," so how these laws are realized varies from state to state, making enforcement, accountability, and understanding on the part of the farmworker more difficult.

Also noteworthy is that the regulation of pesticides as they relate to farmworker safety and health is not covered in OSHA. Rather, oversight of the regulation was given to the Environmental Protection Agency (EPA) due to their expertise in toxicology and environmental assessments. According to Ellen S. Greenstone (1975, p. 72) who wrote on this matter only a few years after OSHA was passed:

> By quietly ceding jurisdiction to formulate pesticide regulations to the Environmental Protection Agency (EPA), an agency which lacks sufficient authority to protect worker interests, OSHA decided that "every working man and woman in the Nation" simply did not include some 2.8 million farmworkers.

Greenstone chronicles the ensuing debate about the inclusion of pesticide regulation in OSHA standards, noting that ultimately the interests of growers as employers won out to keep jurisdiction in the EPA. While OSHA later contained provisions made to ensure that farmworkers were notified of pesticide use, this did little to further protect them because pesticides regulated under the EPA were exempt from their regulations. According to the U.S. Department of Labor:

> Chemicals must be properly labeled so farmworkers know the identity and hazards of the chemicals they may be exposed to at work. OSHA has information to assist employers and workers ensure that hazard

> communication is properly addressed in their workplaces. In addition, certain OSHA standards address hazard communications.
>
> (OSHA, n.d.b)

They go on to acknowledge that pesticides covered under the Federal Insecticide, Fungicide, and Rodenticide Act (FIFRA) are exempt from OSHA regulation because they are overseen by the EPA. So, while hazard communication is part of broader OSHA legislation, it does little to protect farmworkers by giving information regarding one of the top factors affecting their safety and health. Further, because the EPA does not regulate workplace environments, improper use and ignorance of the danger of pesticides are technically outside of its jurisdiction, allowing farmworkers to fall through regulatory cracks when it comes to pesticide use. As Jill Lindsey Harrison (2011, p. 87) points out, regarding pesticide-related policy:

> The story of early pesticide regulatory priorities is not simply one of corrupt bureaucrats and interest group politics. Trends in pesticide policy and regulatory practice must be understood as shaped by the predominant social beliefs and issues of the day.

Pesticide policy is fragmented, resulting in the problem of worker exposure, even when legal pesticides are used properly by the letter of the law.

In 1983 (with amendments in 1986 and 1995), the Migrant and Seasonal Worker Protection Act (MSPA) created housing, employment, and transportation standards, but these standards did not apply to H-2A workers (Wiggins, 2009). The MSPA "protects migrant and seasonal agricultural workers by establishing employment standards related to wages, housing, transportation, disclosures and recordkeeping." It "also requires farm labor contractors to register with the U.S. Department of Labor (DOL)" (U.S. Department of Labor, 2008). The Act provides a series of comprehensive protections to farmworkers, and places the responsibility of record-keeping on the farm employer and/or contractor. However, there are several exemptions to these regulations. They are: (1) the family business exemption; (2) the small business exemption in accordance with the definitions from the Fair Labor Standards Act of 1938; (3) the farm labor contractor exemption, which exempts contractors from registering with the Department of Labor for reasons including being an agricultural employer him/herself, contracting less than 13 weeks per year, or contracting those representing labor organizations; and (4) the farm labor information exemption, which allows certain employers and contractors exemptions from providing information on wages, hours, and working conditions to their employees if they are also exempt in other areas (Harris, 2005). Because of these exemptions and the exemption of H-2A and undocumented workers, this legislation, while important and fundamental for the protection and justice of farmworkers, is insufficient in doing so for a large proportion of foreign-born farmworkers in the United States.

Policy impacting pesticide use, information, and safety

As explored previously, the regulation of pesticides and safety practices regarding pesticides is the jurisdiction of the Environmental Protection Agency (EPA). Pesticide use and regulation have a complicated history between what is best health-wise for farmworkers and consumers when it comes to these chemicals, the need to produce ever-increasing amounts of food, and the perception that pesticide use makes doing so easier. As Jill Lindsey Harrison points out:

> The pesticide paradigm functions ideologically, since this set of beliefs reinforces existing institutions and practices. Regulatory agencies' failure to address the environmental or human health consequences of agricultural pesticides therefore has had as much to do with widespread social beliefs.
>
> (2011, p. 87)

In 1992, the EPA passed the Worker Protection Standard (WPS) with the understanding that full implementation of the regulations it contained would be implemented by 1994–1995. This standard mandated several things, including the provision of training for farmworkers who are working with pesticides. Arcury, Quandt, Cravey, Elmore, and Russell (2001, p. 488) acknowledge that this training is most effective when the farmer or farm owner takes an active role and provides, "(1) safety equipment and sanitation facilities, (2) safety behavior education and training (e.g., how and why to use equipment and facilities), and (3) a work environment conducive to safety behavior (e.g., employer encouragement to work safely)." The WPS requires training for a farmworker who enters an area within 30 days of its treatment with pesticides or one that requires a restricted-entry interval, or a certain amount of time to pass before one can enter without protective equipment. It mandates additional training for workers who labor in enclosed spaces such as greenhouses, and those who work with the original mixing and/or application of the pesticides. These trainings cover: general information regarding pesticides including where they might be encountered during work; the routes, hazards, effects, and symptoms of direct exposure, drift, and residue; emergency care and access to medical care; and reduction of risk by requiring warnings, signage with entry restrictions in treated areas, information about the pesticides being used, and protection for workers that want to report a violation. They must be conducted in a language that is understood by the farmworker. Enforcement of all of these requirements is handled at the state level (Arcury et al., 1999; Arcury et al., 2001).

Despite these protections, in a 1999 study conducted with farmworkers throughout North Carolina, researchers found that "only 35.2 percent … of farmworkers reported ever having received any information or training on pesticide safety" (Arcury et al., 1999, p. 461). Another study that consisted of interviews with farmworkers in North Carolina found, "among farmworkers there is some variability in discussions of workplace safety and sanitation, but most workers feel basic safety and sanitation facilities are not available to them in North Carolina" (Arcury et al., 2001, p. 491). This study also found wide variability in understanding, perceptions, and ability to provide safety and sanitation between farmers, farmworkers, healthcare providers, and other stakeholders related both to their unique vantage point as well as to other cultural and societal factors and health beliefs. They present data that show:

> It is apparent that farmers also have stereotypes about Latino workers and about safety regulations that affect their willingness to meet safety and sanitation requirements. Farmers feel they are over-regulated, and that while they try to follow these regulations, farmworkers disregard their efforts.
>
> (ibid., p. 496)

While both of these studies were conducted in North Carolina, it is likely that the same obstacles to understanding, enforcement, and protection exist in other geographical areas as well. Annual funding and congressional disagreements further complicate the broader limits of the WPS regulation. These issues create obstacles that are likely exacerbated when there are language and other cultural differences between the farmer/farm owner and the farmworkers themselves.

Given these problems and with tireless advocacy of different organizations on behalf of and involving farmworkers, the EPA worked to revise the WPS throughout 2014, with the final rule passing in May 2015. After releasing a draft of the revised standards in 2014, the Farmworker Association of Florida (FWAF) and the Farmworker Support Committee (CATA) conducted outreach so that farmworker experiences relating to pesticides could be submitted as part of the public comments and response. In doing so, "FWAF and CATA ensured that the geographically, socially, and economically isolated farmworker populations in their regions would not continue to be ignored and left vulnerable to pesticide exposure" (Brennan et al., 2015, p. 2). In addition to all of the other comments submitted to the EPA for consideration in the revision of these standards, they included the stories of farmworkers themselves, giving them an opportunity not only to talk about the insufficiency of the current WPS, but also to suggest real options for strengthening it.

The final revised WPS passed in May 2015 states:

> EPA is finalizing updates and revisions to the existing worker protection regulation for pesticides. This final rule will enhance the protections provided to agricultural workers, pesticide handlers, and other persons under the Worker Protection Standard (WPS) by strengthening elements of the existing regulation, such as training, notification, pesticide safety and hazard communication information, use of personal protective equipment, and the providing of supplies for routine washing and emergency decontamination. EPA expects this final rule to prevent unreasonable adverse effects from exposure to pesticides among agricultural workers and pesticide handlers, vulnerable groups (such as minority and low-income populations, child

> farmworkers, and farmworker families) and other persons who may be on or near agricultural establishments, and to mitigate exposures that do occur. In order to reduce compliance burdens for family-owned farms, in the final rule EPA has expanded the existing definition of "immediate family" and continued the existing exemption from many provisions of the WPS for owners and members of their immediate families.
>
> (Federal Register, n.d.)

The proposed changes increase the requirements for certification for those individuals responsible for applying pesticides, and the National Association of State Department of Agriculture (NASDA) opposes these changes, stating:

> The proposed rule incentivizes both the state regulatory agencies and the regulated community to respond to implementation and compliance requirements in a manner that is in direct conflict with the Agency's stated objectives for publishing this proposed rulemaking.
>
> (NASDA, n.d.)

Changes to the WPS in relation to applicator regulation without additional funding are untenable in most states and territories, contrary to research presented by the EPA that states otherwise. This ongoing tension highlights the complex dynamics between farmworker interests and those of government and business/farm owners. The 2016 annual funding bill for the EPA received specific backlash due to the proposed addition of a third party clause to WPS rules. Under the addition, farmworkers and third party representatives would be able to request pesticide information from farmers. While there is speculation about the EPA's intentions behind taking out this clause for congressional approval in 2015 and writing it back in after public comments, it is clear that political alliances and agricultural exceptionalism shape the WPS program due to the EPA's reliance on Congress for funding (Schultz, 2016). Even with the EPA's intentions to create stricter regulations to assist farmworkers, implementation remains reliant on the relevant funding.

Policies impacting immigrants in U.S. agriculture

As reviewed in Chapter 2, 72 percent of farmworkers are foreign-born, with a minimum of 68 percent of all farmworkers coming from Mexico. As such, considering immigration and immigration policy, as well as the landscape of trade policy that creates conditions in other countries that force people to seek work elsewhere, as their very livelihoods are eliminated due to the availability of cheap goods and services in their markets is essential to understanding the landscape of injustice for farmworkers. In this section, I briefly review more recent immigration policy and its historical origins, as well as the North American Free Trade Agreement (NAFTA) and its intersection with environmental justice, in addition to currently contested trade policy. All of this gives a perspective on the complicated web within which farmworkers find themselves exploited but with few other options.

The Bracero Program and H-2A

Illegal immigration is rooted in U.S. agriculture, which has historically had a surplus of land and capital but a lack of labor needed to cultivate it. In his book called *They Saved the Crops: Labor, Landscape, and the Struggle over Industrial Farming in Bracero-Era California*, Don Mitchell (2012) explains that the landscape of farming changed significantly in the early 1900s: increased irrigation reliance resulted in the corresponding capitalization of farms. Average farm sizes decreased while large landowners remained powerful. Most importantly, there arose, along with WWII, "an increased demand for temporary and seasonal labor, labor cheap enough to offset the advantages of mechanized wheat farming, the increasing capitalization attendant upon intensification, and the high transportation costs to distant markets" (ibid., p. 16).

During this time of war, growers were faced with increased labor shortages across sectors, and used their worries to appeal to:

> the federal government to create a legal mechanism to bring Mexican workers into the United States to labor in agriculture ... Without public debate or legislation, a bilateral agreement between Mexico and the United States was signed: Mexican workers would be recruited by U.S. government agents and then sent north to labor for growers who were required to feed, house, and employ them for a set period of time.
>
> (Rothenberg, 1998, p. 218)

These dynamics set the stage to increase the size of the cheap labor force, which came exclusively from Mexico (Carrasquillo, 2012). As Holmes states:

> The reproduction of a system of migrant labor hinges on the inability of the migrants, as individuals or as a group, to influence the institutions that subordinate them to the other fractions of the labor force and to the employer. Ever renewed and updated legal, political, and symbolic separations produce the maximal extraction of labor as well as the inherent suffering and danger linking one side of the border to the other.
>
> (2013, p. 12)

Toward this end, Hanson and Spilimbergo (2001, p. 619) assert, "The stated justification of allowing guest farm workers into the United States has been to prevent labour shortages on U.S. farms. The most well-known guest-farm-worker arrangement was the Bracero Program of 1942–65." Given these dynamics, the U.S. government established the Bracero Program in 1942 to help Mexicans enter the U.S. labor market under conditions of wartime shortages, and this program:

> granted Mexican laborers temporary permits to work in U.S. agriculture. Laborers in Mexico, or known as braceros, were contracted by U.S. agricultural growers to work in the United States for a single growing season, after which they were obliged to return to Mexico.
>
> (Hanson et al., 2001, p. 32)

It is crucial to understand U.S. history and policy surrounding immigrants working in agriculture, as Mitchell (2012, p. 3) puts it, "How the specific structures and practices governing any 'guestworker' program must be understood if we are to develop a clearer understanding of how they operate in general within the overarching practices of labor exploitation in capitalism."

The program clearly increased the number of workers coming to the United States from Mexico, and after the end of the war resulted in, "an oversupply of workers, which in turn essentially eliminated the ability of workers to organize" (Allen, 2004, p. 31). As Allen further points out, because the federal government covered many of the expenses related to bringing these workers to the United States and back to Mexico, "the agricultural industry, primarily large farmers, received a labor subsidy in the form of cheaper immigrant labor not available to any other American industry" (ibid., p. 136). According to the Bracero History Archive, "From 1942 to 1964, 4.6 million contracts were signed, with many individuals returning several times on different contracts, making it the largest U.S. contract labor program" (Bracero History Archive, n.d.). The Bracero program ended in 1964 "because of vigorous opposition from labor unions, religious organizations, and community groups" (Rothenberg, 1998, p. 219).

Understanding the Bracero program provides an important foundation for current guestworker policies, which were formulated in the late 1990s. In agriculture specifically, "growers frequently declare that without low-wage foreign labour they would be forced to cut back production or shut down their operations altogether" (Hanson & Spilimbergo, 2001, p. 616). Changes in prices and incentives in the United States can be a motivator for corporate agriculture to band together to form lobbies, pulling for the government to loosen restrictions on the entry of undocumented immigrants into the country.

After the end of the Bracero program, it was replaced with its modern-day counterpart guest worker program, the H-2 (or H-2A) program. This program in theory provides protections for domestic workers. According to Rothenberg (1998, p. 218), "Foreign workers were only supposed to be recruited for jobs where no domestic workers could be found, and the wages and working conditions for H-2 workers were not supposed to depress the local market." The Department of Labor would theoretically only approve requests where growers could prove that they had adequately searched for laborers within the domestic market, and where they could provide housing, contracts, and fair wages. On the Department of Labor's site it states, "The H-2A temporary agricultural program allows agricultural employers who anticipate a shortage of domestic workers to bring nonimmigrant foreign workers to the U.S. to perform agricultural labor or services of a temporary or seasonal nature" (Department of

Labor, n.d.). When examining the reality of justice as it pertains to workers here through the H-2A program, Gray (2014, p. 52) illuminates:

> Legally in the country on work visas that were tied to a specific workplace, they were not able to terminate employment with the same ease as undocumented workers, and their retention in contracted work largely depended on their employers' positive assessments. Consequently, they had less incentive to complain about their work situation.

In actuality, this is yet another program that allows for the potential exploitation of workers from other countries. Cindy Hahamovitch puts it in relation to the Atlantic Coast, "The H2 program gave the nation's growers the power to seek labor from abroad at taxpayers' expense, Atlantic Coast growers began to employ Western methods of labor control" (1997, p. 201), which not only served to diversify the labor force, but to create a system in which different migrant groups were in competition with one another for work. This led people to accept things they might not otherwise so they would get jobs. While there are some policies that protect H-2A workers such as housing regulations and inspection, workers' compensation coverage, and minimum work guarantees, as seen throughout this chapter and in Chapter 2, these workers are many times excluded from protective policies, as well as not being accounted for in data sources such as the National Agricultural Workers Survey (NAWS). While there is a great deal of research surrounding these guest workers, it is difficult to get a complete sense of their complex and nuanced realities.

The Immigration Reform and Control Act

Given the number of farmworkers who are not born in the United States, policies that relate to immigration directly affect a large proportion of individuals laboring on U.S. farms. As Brown and Getz (2011, p. 136) put it, the "significance of border and immigration politics in mobilizing anti-immigrant sentiment and undermining bargaining power for farmworkers cannot be overstated." They illuminate the dynamics surrounding immigration control and enforcement, as well as the centrality of policies pertaining to immigration and their impact on the maintenance of an exploitable class of workers. The Immigration Reform and Control Act (IRCA), passed during the Reagan administration (1986), allocated more resources to control our borders and "sought to reduce the presence of unauthorized immigrants in the United States by allowing agricultural workers to apply for legal status, legalizing approximately 1.1 million farmworkers, who then moved out of agriculture and were replaced by new undocumented entrants" (ibid., p. 136). Other sources indicate that closer to 2.7 million agricultural workers gained legal status after this Act was passed (Rytina, 2002; Gray, 2014). This Act was an amendment to the Immigration and Nationality Act of 1952 (Harris, 2005).

Durand, Massey, and Parrado (1999) also discuss IRCA, explaining that in this Act were two provisions offering amnesty and legalization to undocumented workers, particularly in agriculture. These provisions and others were made to gain the support necessary from growers and activist groups to pass the policy. While one of the aims of the Act was to curb undocumented immigration to the country, it simply made it easier for workers to gain citizenship, transforming seasonal workers into more permanent residents. For this reason, this Act set the stage for and was pivotal in an uptake of Mexican immigrants to the United States, with which came many human rights and social justice side effects. Discrimination by employers toward Latinos and Mexicans began to run rampant, and most of this took the form of lower and insufficient wages. In addition to these statistics, Durand, Massey, and Parrado present a compelling case that after time is spent working in the United States, the likelihood of an immigrant to try to bring his/her family into the United States increases exponentially. Gray points out in her discussion of IRCA that it "included the Seasonal Agricultural Worker (SAW) provisions, which offered a path to citizenship for anyone who could prove they worked at least ninety days as a farmworker in the three years prior to the act's passage" (2014, p. 114) and that this subsequently encouraged further undocumented immigration, as well as putting those who had earned amnesty in a position to recruit workers from their home communities.

AgJOBS and the DREAM Act

Aside from policies regarding workers' rights and the number of immigrants working in agriculture, there are a number of other policies that could affect non-U.S.-born farmworkers. Among the most hotly contested of these are the DREAM (Development, Relief, and Education for Alien Minors) Act and AgJOBS (Agricultural Job Opportunities, Benefits, and Security) Act. Both Acts have had bipartisan support, but after many years of congressional debates have never passed into law.

The DREAM Act (DREAM Act, n.d.) focuses on a citizenship track for minors born outside of, but raised in, the United States. These minors, many unaware that they are not citizens due to complicated family situations, have historically been threatened with deportation and lack of aid and opportunity once they graduate high school and apply for college. Many children of immigrant farmworkers fall into this category, and, as such, this legislation draws support from organizations advocating farmworker rights.

One of the longest-debated, most sweeping and well-known pieces of legislation in the United States regarding immigrants working in U.S. agriculture is called the Agricultural Job Opportunities, Benefits and Security Act (AgJOBS). It deals with a fast track to citizenship for agricultural workers who are here working through the current guestworker (H-2A) program, in addition to proposing changes to the current H-2A program. AgJOBS has been on the floor in Congress for debate intermittently since 2000. According to the Farmworker

Justice website (Farmworker Justice, n.d.), "AgJOBS, the Agricultural Job Opportunities, Benefits and Security Act, is a proposed immigration law that would provide agricultural employers with a stable, legal labor force while protecting farmworkers from exploitative working conditions."

> If enacted, AgJOBS would (1) create an "earned adjustment" program, allowing many undocumented farmworkers and agricultural guestworkers to obtain temporary immigration status based on past work experience with the possibility of becoming permanent residents through continued agricultural work, and (2) revise the existing agricultural guestworker program, known as the "H-2A temporary foreign agricultural worker program."

While AgJOBS was first proposed in 2000, it has never been passed in any of its iterations since then. In that time, farmworker advocates have advocated for revisions to this legislation to ensure more protections for foreign workers, along with simpler procedures for becoming U.S. citizens if they desire.[4]

Trade agreements

Our preconceptions of the visible and invisible work of immigrants have implications for work in the field of justice and the political economy of undocumented immigration in the United States. One could argue that our various free trade agreements, paving the way for cheap goods and services into the United States, were merely catalysts bringing to life reactions about immigration that were brewing for quite some time. Illegal or legal, undocumented or documented, large corporations in many sectors of the economy exploit and use for profit the plight of immigrants. The answers to undocumented immigration require more than responding to "How do we stop it?" Instead, questions should deeply consider, first, the conditions that create and sometimes even encourage undocumented immigration and, second, consider that *if* some industries cannot exist without this labor, the labor itself at least needs to be dignified. Political and public recognition that the nuances associated with going beyond surface solutions towards broad reform that would treat the underlying problem are key in envisioning a new way for neighboring countries to peacefully co-exist with one another. Further, what are the implications for environmental justice to these web-like, complex dynamics?

One of the biggest considerations in the absorption of undocumented workers in the United States is its effect on wages for domestic workers. Howell (2007) demonstrates that during immigrant surges (such as post-NAFTA to present), there has been a decrease in wages for immigrant workers, which can in turn decrease wages for legalized foreign-born workers and low-educated African Americans. While employment opportunities and wages fall, it follows that anti-immigration sentiments have risen. Sources of these sentiments of anger vary (fiscal setbacks, violations of national and local laws, racism, etc.). Crucial to this view are the concepts of increasingly lower wages and worsening labor conditions due to the surge of low-skill immigration, resulting in "very low reservation wages (the wage at which workers will agree to offer their services) and no bargaining power" (ibid., pp. 2–3).

Thus, while folding undocumented workers into the workforce in sectors like agriculture benefits the employers and the consumers, often those who would work in similar sectors not only lose their jobs, but also lose bargaining power. In addition to a growing anti-immigrant sentiment, particularly during periods of economic slump, we see a relaxation of enforcement in worker protection policies in industries that are large economic contributors. Agriculture in particular has a great deal of lobbying power, and as such, the standards for immigration and workers' rights set forth by the United States Immigration Reform and Control Act are not always enforced. Also, agriculture is the only sector where officials actually need a warrant to search (Hillman & Weiss, 1999). Farm labor contractors hire approximately 185,000 undocumented immigrants as laborers each year, making up, as previously stated, 30 percent of all agricultural workers in the United States. In agriculture and other influential sectors, "permissible illegal immigration" is supported by institutional policy and a "don't-ask-don't-tell" mentality, while those who are losing jobs to these workers or making less money due to their presence become increasingly frustrated.

NAFTA, which created a displacement of peasant farmers in Mexico and increased economic prosperity in the United States, has led to a decrease in the militaristic nature of our borders (Andreas, 1998). It is easier to get across the borders, and since the passage of NAFTA, there has been more porous crossing along the U.S.-Mexico border. But, as Deborah Barndt points out:

> While the journey of the tomato from the Mexican field to Canadian supermarket is framed, historically and geographically, by the North American Free Trade Agreement, NAFTA is merely a continuation of policies and practices that have long been shaping the global economy.
> (2002, p. 68)

In this same continuation of policies, Martinez (1993) reminds us that human rights were also not a consideration as legislators negotiated and drafted NAFTA. In addition, during the same time the militaristic control of the border was increasing under the pretense of keeping drugs from flowing into the United States at the very same time NAFTA was being drafted, ensuring the economic integration of the two countries. Finally, he points out that "the dangers of illegal border crossings is one of the reasons why NAFTA should contain specific provisions for protecting the labor, human, and civil rights of immigrants" (ibid., p. 3).

One distinction that Raul Delgado-Wise (2004) makes is that Mexico's "export-led" economy rests heavily on the export of not only manufactured goods, but also people. Exporting the labor force is thus masked as 'progress' while diminishing Mexico's ability to truly develop as anything more than just a reserve of labor for capital elsewhere. Delgado-Wise helpfully discusses some of the qualitative implications of these numbers. He states:

> Along with these characteristics, there has also been a significant increase in the flow of remittances sent to Mexico, which increased *fivefold over the past decade* to reach a historical maximum of US $13,266 million in 2003. Not only does that consolidate Mexico's position as the world's main recipient of remittances or "migra-dollars" alongside India, but it also means that exports of labour are the country's third most important source of foreign earnings, with a contribution to the balance of payments that exceeds those made by tourism and agricultural exports.
>
> (ibid., pp. 595–596)

According to Brown and Getz (2011), remittances account for around 3 percent of Mexico's Gross Domestic Product (GDP), and in areas like Oaxaca and Chiapas, remittances account for upward of 50 percent of income.

Thus, the United States must consider the following three policy questions, each of which has implications from an environmental justice perspective, as highlighted by Hanson et al. (2001):

1. What are and what should be the level and composition of *legal* immigration allowed in the United States?
2. What do we do about undocumented immigrants, especially considering that they account for one-third of all new immigrants to the United States each year?
3. Should immigrants (documented and undocumented) be able to apply for public assistance and/or should there be an easier path to citizenship?

Haus (1995) further considers the implications of these three questions on the notion of state sovereignty. Because the consequences of migration correlate with the consequences of trade and foreign investment (i.e. NAFTA), what is/should be the role of the state in the control of the coming and going of immigrants? There is no protocol for considering these intersections of political, economic, and justice-related issues centered on the inherent debates around undocumented immigration and its place in American society.

To further understand the social dimensions of farm labor in the United States, it is useful to have a brief history of the political and economic context of immigrants working in agriculture in the country and how this has been affected by free trade policy. While it was President George H.W. Bush who first researched, proposed, and sought to implement NAFTA, it was President Bill Clinton who signed it into law in December of 1993. Once implemented on January 1, 1994, there was an immediate lifting of tariffs on one-third to one-half of the imports from Mexico to the United States and exports from the United States to Mexico. In the years after, further elimination of trade tariffs 'freed' trade between the United States and Mexico (Canada is not mentioned here due to the fact that the United States–Canada trade was predominately duty-free because of the 1988 Canada–United States Free Trade Agreement) with effects not only on trade, but also the environment, industry, and people. As Douglas Massey and Kristin Espinosa (1997, pp. 991–992) explain:

> The provisions of NAFTA … help to bring about the social and economic transformations that generate immigrants. The integration of the North American market will also create new links of transportation, telecommunication, and interpersonal acquaintance, connections that are necessary for the efficient movement of goods, information, and capital, but which also encourage and promote the movement of people – students, business executives, tourists, and, ultimately, undocumented workers.

As explained previously and supported by environmental justice scholars like David Schlosberg (2004; 2007) and John Callewaert (2002), such historical systemic understanding is needed to expand and strengthen the field. In *The High Price of Free Trade*, Robert Scott asserts that NAFTA was designed to provide investors the guarantees necessary to drive foreign direct investment and transition factories from within North America and the hemisphere in general. The core focus contained no protections to labor or environmental standards maintenance (2003, p. 11).

The North American Free Trade Agreement has reduced the number of jobs available in the United States, contributed to rising income inequality, suppressed real wages for production workers, weakened workers' collective bargaining powers and ability to organize unions, and reduced fringe benefits (ibid.). While this has affected many industries, it has had a dramatic effect on agriculture. At a time in U.S. history when conventional agriculture and the mass-production of food are becoming more commoditized, the United States passed a free trade agreement that made it simpler for agricultural corporations to hire and exploit immigrants as a cheap, unskilled labor force.

As noted by Peter Andreas (1998), NAFTA, while predicted to eventually curb illegal immigration to the United States, set up the conditions for a short- and medium-term spike in the number of immigrants remaining undocumented in the United States. As he further points out, "since the late 1980s, the Mexican government has been cutting back electricity, fertilizer, water, and credit subsidies to peasant farmers," and other subsidy-related cutbacks to affordable credit, infrastructure related to irrigation and drainage, and crop insurance that are "needed to modernize Mexican agriculture and make small-scale farming truly viable" (ibid., p. 609). Holmes further elucidates:

> Thus, the Mexican government was forced to erase tariffs, including that on corn, the primary crop produced by indigenous families in southern Mexico. However, NAFTA and other free trade policies do not ban government subsidies. Thus the U.S. government was allowed to increase corn subsidies year after year, effectively enacting an inverse tariff against

> Mexican corn. In addition, such subsidies are only possible for relatively wealthy countries and could not be enacted by the relatively poor Mexican government.
>
> (2013 , p. 25)

Due to these cutbacks, Mexican small-scale communal farming, which had sustained its rural population for generations, is not viable an option. The actions taken by the Mexican government in the lead-up to NAFTA, in combination with the devaluation of the Mexican peso after its enactment, ensured a steady supply of people with no other option than to come to the United States to seek menial work.

In the 1990s, while many industries applauded the effects of NAFTA in liberalizing trade and the exchange of capital, they were less enthusiastic about the levels to which NAFTA had increased the flow of labor. Hanson et al. (2001) point out that this led industries to lobby for more work visas, claiming that there was a shortage of domestic labor willing to do the work necessary to meet production and supply needs, which were increased dramatically by NAFTA as we flooded our cheap goods into other markets. With these conditions as a foundation, "current policy appears to favor individuals with relatively low levels of education and other observable skills" (ibid., p. 3).

Raul Delgado-Wise (2004) points out that immigration from Mexico to the United States over the decade following NAFTA was over ten times what it was in the 20-year period prior to its passage. In addition to this, it is clear that within this time not only did the numbers of immigrants dramatically increase, but they diversified in geographical origin, job sectors, age, race, and gender, as well as how and when remittances to the home country are sent and their impact. The implications of these assertions are multifaceted, but one thing is for certain; the amount of remittances, or money sent back to Mexico by laborers in the United States, increased fivefold in the decade following NAFTA's passage. The effect of this on the Mexican economy is that exported labor now serves as the country's third most important source of foreign earnings. The effects of NAFTA both on the viability of the Mexican economy, its reliance upon U.S.-driven solutions, and a marginalized group of people have made the nearly 20 years following NAFTA a time of great change, and often desperation, for Mexico.

Broader effects of these dynamics are felt in different sectors of the central U.S. economy as the favorability of immigration waxes and wanes, and correlates also with the intensity of border enforcement. As argued by Hanson and Spilimbergo (2001) border enforcement is relaxed when the U.S. economy is strong and there is a high need for cheap labor in menial labor sectors such as agriculture and factory work. They point out that approximately 300,000 new illegal immigrants enter the United States each year, and are an important part of "agriculture, apparel, construction, food processing, lodging, restaurant, and domestic help industries" (ibid., p. 613). California, Florida, and Texas have the highest concentrations of undocumented immigrant workers in the United

States, which is likely, according to the authors, correlated to the power of the agricultural industry in these states. Immigration services frequently ignore monitoring of many of the large-scale production locations of these corporations to avoid "sectoral shock" or a negative impact on the ability of these companies to do "business as usual."

It is important to note that most ideas and models used to understand immigration focus on the effects on the host country, rather than the effects of larger economic and trade policy like NAFTA on the source country, in this case, Mexico. George Borjas (1999) points out that without exploring *why* people immigrate in the first place due to economic and political pressures in their home countries, which are ironically many times created or at least fueled by the host countries, it is impossible to recognize and alleviate the pressures caused by immigration and to find ways for labor to flow more effectively across the market. In the case of agriculture, many Mexican subsistence farmers were unable to make a living post-NAFTA due to the flow of cheaper goods into their markets, leaving these farmers and workers looking for jobs that no longer existed in Mexico (ibid.). Adding to this, Sandy Brown and Christy Getz (2011) blame neoliberal domestic policies and international trade for boosting corporate agribusiness while forcing Mexican farmers off their land, resulting in increased numbers of farmworkers from rural Mexico post-NAFTA.

Borjas (1999) urges those looking at the complex dynamics of immigration to look not only at the effects of it on the "host" countries, but also on the "source" countries as well. He writes:

> Immigration also has economic effects on the host country that extend far beyond the labor market … An important part of the modern debate, for instance, concerns the impact of immigrants on expenditures in the programs that make up the welfare state.
>
> (ibid., p. 1699)

He further emphasizes that when the source country creates a disincentive through taxation and removal of support structures, the host country's call to tax the highly skilled worker and to subsidize the less skilled is a sign to employers of workers in low-skill jobs to recruit from Mexico and leaves the low-skilled Mexicans no other option than to comply.

Building upon these ideas, David Howell (2007) argues that the subsequent large long-term immigrant surges put in motion by these dynamics have at least small negative wage effects for less-advantaged members of the labor force. Howell demonstrates that these less-advantaged members are likely to be largest for earlier cohorts of foreign-born workers and less-educated workers in major immigrant-receiving regions. During immigrant surges (such as post-NAFTA to present) there is a decrease in wages for immigrant workers, which can in-turn decrease wages for legalized foreign-born workers and less-educated workers. As Howell points out, these dynamics have implications elsewhere and he suggests that justice for undocumented and other immigrant workers comes at the hands of three considerations: (1) fair wages; (2) employment opportunities (opportunities for advancement); and (3) the development, regulation, and oversight to ensure humane working conditions. These considerations are the very items that, according to Scott (2003), NAFTA's effects have suppressed and made difficult to achieve.

With even this preliminary understanding of one of our longest-standing free trade agreements, we can begin to see how similar agreements with other countries, particularly developing countries, may have similar effects. The United States has free trade agreements with 20 countries, as noted by the United States Department of Commerce. In addition to NAFTA, its expansion, known as the Central American Free Trade Agreement (CAFTA) was controversial in its passage among those working on immigrant farmworkers' rights. It was implemented in its preliminary stages in 2004 with Costa Rica, the Dominican Republic, El Salvador, Guatemala, Honduras, and Nicaragua. Since this time, U.S. agriculture has seen similar surges in workers from these areas, as it did in Mexico after the passage of NAFTA (International Trade Administration, n.d.).

Eytan Meyers (2000, p. 1258) argues that, "The interest and pressure groups most commonly associated with immigration policy are employers and ethnic groups, which tend to support immigration, and unions and nationalist groups, which tend to oppose it." More so, the implications of the nation-state, i.e. "pursuit of economic security, a rational bureaucratic agenda, and continued legitimacy," stand at a time in contrast with its interest in "various societal elements – labor, capital, ethnic groups, humanitarian organizations, and the provinces" (ibid., p. 1261) in the formation of immigration policy. Kenneth Roberts (1997) emphasizes the power of corporate agriculture by inferring a correlation between increased profits in agriculture with increased industrialization of agriculture—leading to more out-of-work, easily exploitable immigrants.

Environmental justice and free trade

How does free trade tie into the discussion of environmental justice, and how might that discussion relate to the dynamics we have unearthed regarding the conditions they face? Amartya Sen (2009) tackles the theoretical issues behind our conceptualizations of justice head-on in *The Idea of Justice*. In this book, he discusses transcendental theory that supposedly represents both individual and collective voice, as well as the concept of choice. He examines the role of power, and the inevitable need to look at the ramifications of actual behavior in real-life situations. He analyzes the inherent difficulties in following the theories of justice to which we default, that, by their very nature, are impractical and inapplicable. Sen also reframes the notion of fairness around the materials of justice by looking at the balance between people's capabilities and their access to things such as life, freedom, resources, happiness, well-being, and liberty. In doing this, he begins to address the complexity of understanding that all people actually *are* created equal with the stark realities of the practical implications of that very statement—insomuch as, if that is true, then all people should have an equal chance (a fair chance) to access those things and live the lives they want. What complicates this is the imposition of social stratification and the history of humans as creatures who have tended toward separation rather than integration, competition rather than collaboration. The parallels between Sen's notions of justice and the conundrums associated with undocumented immigrants serving in U.S. agriculture are many.

As reviewed in Chapter 1, when expanding the scale of environmental justice considerations, Schlosberg states:

> Certainly, at the centre of the recent protests against global financial and trade institutions, against the globalisation of the food system, and for indigenous rights, is the issue of equity; economic or distributive injustice is a key and constant rallying cry. The most basic critique is that the currently favoured model of development increases and exacerbates inequity, both between the North and the South and between elites and the impoverished in southern nations – again, not just in economic goods, but in environmental goods and bads as well.
>
> (2004, p. 523)

When balancing notions of environmental justice, not to take these distributive and participatory elements into account is, in Schlosberg's opinion, the downfall of the environmental justice movement. As such, work surrounding the rights of undocumented immigrants working in agriculture as refugees of the impacts of free trade in their home countries must consider these dimensions. He notes as well a lack of connection between the environmental justice movement and other movements centered around related but divergent notions of justice: economic, political, social, etc. He calls upon Vandana Shiva's (2005) work on the globalization of food production as an example of true linkages between ecological, environmental, sociological, economic, and cultural systems.

Schlosberg (2007) also discusses theorists who look not only upon distribution as the means toward justice and equality, but also look toward the causes of the present misdistribution of goods, services, and harms. Particularly as connected to undocumented immigrants working in U.S. agriculture, he explores the multifaceted nature of just what people consider "environmental justice," ranging from specific toxics issues all the way to deeper issues of race and inequality. This discussion sheds light on why people are often confused about the concept of environmental justice and find it somewhat nebulous in nature, and he suggests seeking the unity that exists among the varieties of environmental justice work.

David Howell (2007) suggests that justice for undocumented workers comes at the hands of three considerations: (1) fair wages; (2) employment opportunities (opportunities for advancement); and (3) humane working conditions. Kenneth Wilson and Alejandro Portes (1980) started with the idea of a mode of formally incorporating undocumented immigrants into the U.S. labor market. They identified this as the foundation and recognition upon which future justice work would be built. Gary Becker (1962) suggests even earlier that a building block to this incorporation is an investment in the education and training of undocumented immigrants.

There is a lack of dignity and opportunity in the work that the undocumented farmworkers do, further amplified by the fact that many of them were brought across the border at a debt to a coyote or corporate contractor. They start their time in our country not only illegally, but many times also in financial obligation to those who bring them here. This is a hopeless situation, and without serious action on some of the above-stated environmental justice recommendations, many of these faces will continue to be lost—unrecognized for their unique contribution to American living, and the sweat that paves the road to the American dream being fulfilled for those with citizenship. Finally, as Douglas Massey (1988) suggests, we need to reinvest the capital gained by the use of undocumented immigration into increasing human productivity, wealth, and national income in the home country of Mexico.

Moving forward, and in light of current conflicts surrounding immigration, much educational work needs to be done in understanding the intricacies of immigration and its economic contributions to our country. It is promising to see so many people protesting what is being done across the country, and if nothing else, perhaps the contributions of undocumented immigrants will be dignified in the process of repeal of current immigration processes and policies. However, one cannot deny the very real threat of backlash, an increase in the militaristic nature of our borders and an increase in the exploitation of immigrants currently residing in the United States. There is no point of intervention that is both necessary and sufficient in itself, and the recognition of the interplay of history, macro-level and micro-level economics and politics, as well as the role of business in justifying the inhumane treatment of these citizens of the world is a start in addressing the problems associated with NAFTA and its role in undocumented immigration—dignifying and recognizing the work of hundreds of thousands of workers in the United States.

Current legislation

Two different pieces of legislation currently on the fast track to passage will impact trade into and out of the United States. Information about the first, the Trans-Pacific Partnership (TPP) includes:

> On December 14, 2009, United States Trade Representative Ron Kirk notified Congress that President Obama intends to enter into negotiations of a regional, Asia-Pacific trade agreement, known as the Trans-Pacific Partnership (TPP) Agreement with the objective of shaping a high-standard, broad-based regional pact. The TPP will help to expand American exports, saving and creating good jobs here at home. The TPP will build upon the United States' existing trade agreements with Australia, Canada, Chile,

> Mexico, Singapore, and Peru and will forge new opportunities to U.S. exports of goods and services to Brunei, Japan, Malaysia, New Zealand, and Vietnam.
>
> (International Trade Administration, n.d.)

The countries included in the TPP, according to Jesse Marx (2016), represent 40 percent of the global economy. While all countries signed the agreement on February 4, 2016, Marx states that the next steps are that, "Congress needs to draft legislation that will make the trade deal enforceable on American soil, possibly as soon as this spring. But if ratification is delayed, the next president could have a say in whether the deal survives" (para. 16).

The second is the Transatlantic Trade and Investment Partnership (T-TIP), "an ambitious, comprehensive, and high-standard trade and investment agreement being negotiated between the United States and the European Union (EU)" (United States Trade Representative, n.d.). In February 2016, the twelfth round of negotiations concerning the T-TIP took place in Brussels, and during this meeting, "Negotiators worked towards the objective of achieving an ambitious, high standard T-TIP agreement that responds to both EU and US interests. They remain ready to seek to conclude our negotiations in 2016, provided that the substance is right" (EC, n.d.).

Since these are currently being debated, one is left to speculate at the consequences of the passage of these agreements on farmworkers, but NAFTA and CAFTA have provided a yardstick by which one can measure potential impact. As stated in a recent Farm Aid blog post:

> Increasingly, corporate-written deals are placing the profits of multinational corporations – who can benefit wherever they choose to set up shop – above the needs of everyday people rooted in their home countries. Taken as a whole, NAFTA-like policies that remove any protections for local workers and farmers put family farmers out of business, send good jobs offshore and trigger a race to the bottom for wages in U.S. working and middle classes.
>
> (Farm Aid, n.d.)

These trade deals do not have the support of labor and environmental groups, who will continue to fight to ensure that they do not get passed. This is ultimately dependent upon the Obama administration's ability to push them through prior to the end of 2016, and if enacted, time will tell their effects on farmworkers and their rights in the United States.

The role of policy in organic agriculture: definitions of organic, certification, and impact on practice

According to the International Federation of Organic Agriculture Movements (IFOAM), "Organic agriculture is a production system that sustains the health of soils, ecosystems and people. It relies on ecological processes, biodiversity and cycles adapted to local conditions, rather than the use of inputs with adverse effects" (IFOAM, n.d.a). They frame organic agriculture as a combination of tradition, innovation, and science, and they mention that it should ultimately benefit the environment and promote fair relationships and a good quality of life for all involved. Another definition of organic agriculture states:

> Organic agriculture is a defined and certified system of agricultural production that seeks to promote and enhance ecosystem health whilst minimizing adverse effects on natural resources. It is seen not just as a modification of existing conventional practices, but as a restructuring of whole farm systems.
>
> (Morison et al., 2005, p. 25)

The U.S. Department of Agriculture (USDA) has developed organic certification standards that privilege ecological and environmental goals but are vague on social and economic goals within the movement. Certification is conducted in a decentralized way by different state-identified entities. The USDA certification language that these state-level entities are to enact is as follows:

> USDA organic products have strict production and labeling requirements. Unless noted below, organic products must meet the following requirements:
>
> 1 Produced without excluded methods, e.g., genetic engineering, ionizing radiation, or sewage sludge.
> 2 Produced per the National List of Allowed and Prohibited Substances (National List).
> 3 Overseen by a USDA National Organic Program-authorized certifying agent, following all USDA organic regulations.
>
> (USDA, 2012)

In addition to considerations concerning the elements included in organic certification, there is the question of whether certification is 'worth it' to the farmer. IFOAM addresses this by discussing,

> There are organic farmers for whom formal certification does not have any advantages: this is true for farmers who practice subsistence farming and do not engage in the market at all, and for farmers for whom the organic claim has little or no marketing value. These groups engage in organic agriculture because of benefits such as increased productivity, lower production costs, a healthier working environment, and other social, environmental, and economic sustainability considerations.
>
> (IFOAM, n.d.b)

This statement highlights the diversity of rationale that informs farmers' decisions to farm organically versus using conventional methods. This makes sense, considering that the roots of the organic movement were in a counterculture that wanted to illustrate how we can redevelop society by redeveloping the food system. This is also consistent with information collected in research regarding organic agriculture about family members providing most of the labor on the farms so any capital gained can be used to pay for necessary expenses (see Chapter 5). This trend and the tendency to equate organic agriculture with family labor have led to a disregard/lack of understanding of the experiences of hired labor within these systems as organic agriculture has grown in size and scope (Shreck et al., 2006).

In *Agrarian Dreams*, Julie Guthman (2004) discusses farmworkers on organic farms, particularly as the organic movement was co-opted to include large-scale farms in addition to family-scale ones. She points out that growers and marketers of the organic movement state that labor is outside the scope of the original philosophy of organic agriculture, except when convenient for their purposes. Regarding the impact of this on consumers' mentality regarding organic options, she states that they "choose organic food with the assumption that it is somehow better for farmworkers," yet, all the while the inclusion of "labor standards never had much of a chance, since one of the organization's highest dues-paying members had been the target of the United Farm Workers' grape boycott" (ibid., p. 119). She also later goes on to point out that the USDA ownership of the organic certification has prompted many farmers into "exiting the formal organic sector altogether and finding other ways to differentiate themselves" (ibid., p. 170).

The organic movement has repeatedly attempted to reframe the negative aspects of conventional synthetic pesticide and fertilizer use (Vogl et al., 2005). More recently, the organic farming movement has itself been the subject of increasing levels of criticism as it adopts more industrialized scales of production. While organic farming began largely as a bottom-up movement originally initiated by farmers themselves, it has moved to a process necessitating certification and public accountability that can be easy for industrial operations to mechanize and difficult for small farmers to navigate. While certification has proven evolutionary in nature, with recent changes including provisions relating to outdoor access and living standards for poultry and livestock, the definition's expansion still fails to address labor in any capacity (Starmer, 2016). In addition, there are not consistent standards across borders.

As organic agriculture has evolved in recent decades, policy that supports it has in many ways lagged behind. The Farm Bill, passed officially as The Farm Security and Rural Investment Act of 2002, expired in 2007 when Congress passed an extension to 2012 and next comes up in 2017–2018. Congress continued debating and refining a new Farm Bill, The Food Conservation and Energy Act of 2008, with a retained focus on revitalizing rural areas as well as new goals:

> building on momentum of the ag industry and rising farm income; contributing rural communities and infrastructure; supporting the bioeconomy;

> protecting nutrition assistance; developing a farm safety net; enhancing conservation and clean energy; promoting markets at home and abroad; and promoting research.
>
> (Baum, n.d.)

While nuanced, the interaction of the farm bill with trade policy as well as the subsidization of certain crops does not bode well for organic agricultural techniques, and since the first legislation was passed, it has not resulted in positive labor changes, with the number of rural agricultural jobs continuing to drop (James & Griswald, 2012).

In early 2014, the new Farm Bill—the Agricultural Act of 2014—was signed into law. As expected, there were some wins for sustainable and organic agriculture. These include investments in beginning farmers, giving them access to land, credit, and training, more funding for research in organic agriculture, provisions making it easier to spend food stamps at local farmers markets, policy ensuring that farmers who receive crop insurance subsidies use natural resources wisely on their farms, and access for farmers with diverse crops and livestock to get insurance tailored to their needs. Also as anticipated, there were some losses as well, some connected to larger losses of public assistance funding, and others specifically affecting farmers of color, rural small business entrepreneurs, the environment (funding for smart resource conservation was cut dramatically), and small and mid-size farmers (there were no subsidy reforms, which means that they remain uncapped and unlimited, which ultimately benefits large, wealthy farms) (NSAC, n.d.).

Conclusion

The conditions explored in Chapter 2 are by and large created by the historical and policy-related elements considered in this chapter. When looking over time at the fate of farmworkers it is notable that little has changed regarding the justice-related issues they face and the larger political-economic conditions within which those issues are created. In many ways, policy has served to reinforce the troubling, paternalistic system within which farmworkers' livelihoods are realized, from finding work, to making their way into the United States, to exposure to substandard and egregious working and living conditions, to lacking a voice in the very processes which shape these dynamics. Now, I will turn to two case studies, one rooted in conventional agricultural and the work of the Farmworker Association of Florida (FWAF) and the other in small- and family-scale organic agriculture in the Northeastern United States in partnership with the Northeast Organic Farming Association. In Chapters 4 and 5, I will show how much of what has been discussed plays out 'on the ground' so to speak. Chapter 6 will move to justice claims surrounding how things should be, the work being done to create those conditions for farmworkers, and more about the impact of fair trade certification on agricultural labor.

Notes

1 There were many policies and movements prior to those covered in this chapter that set the stage for what is explored herein. For a succinct and comprehensive review see the timeline page of National Farm Worker Ministry website at: http://nfwm.org/education-center/farm-worker-issues/timeline-of-agricultural-labor/ or review their slides on some of the historical dynamics of farm labor at http://nfwm.org/wp-content/uploads/2014/02/immigration-timeline-FINAL-6.25.13.pdf

2 Note that many state-level policies impact farmworkers as well, particularly as wage and fair labor standards apply and are enacted. See Arcury and Quandt (2009) for a review of these policies in the states of the Eastern United States.

3 For more information on the hazards that farmworkers are exposed to and their various protections according to law, see the United States Department of Labor page at: www.osha.gov/dsg/topics/agriculturaloperations/hazards_controls.html

4 For more information on immigration reform and farmworkers, including other legislation that would reform the H-2A program, see the Farmworker Justice website at: www.farmworkerjustice.org/advocacy-and-programs/agjobs. In addition, they provide information on the most recent legislative action passed by Obama here: www.farmworkerjustice.org/sites/default/files/President%20Obama%20s%20Executive%20Action%20on%20Immigration.pdf that relates to the purposes of the DREAM Act and AgJOBS.

References

Allen, P. (2004). *Together at the table: sustainability and sustenance in the American agrifood system.* University Park, PA: Pennsylvania University Press.

Anderson, P. M. (1988). Agricultural Employee Exemption from the Fair Labor Standards Act of 1938. *The Hamline Law Review*, 12, 649.

Andreas, P. (1998). The Escalation of U.S. Immigration Control in the Post-NAFTA Era. *Political Science Quarterly*, 113(4), 591–615.

Arcury, T. A., & Quandt, S. A. (2009). *Latino farmworkers in the Eastern United States health, safety and justice.* New York: Springer. Available at: http://public.eblib.com/choice/publicfullrecord.aspx?p=428954

Arcury, T. A., Quandt, S. A., Austin, C. K., Preisser, J., & Cabrera, L. F. (1999). Implementation of EPA's Worker Protection Standard Training for Agricultural Laborers: An Evaluation Using North Carolina Data. *Public Health Reports (1974-)*, 114(5), 459–468.

Arcury, T. A., Quandt, S. A., Cravey, A. J., Elmore, R. C., & Russell, G. B. (2001). Farmworker Reports of Pesticide Safety and Sanitation in the Work Environment. *American Journal of Industrial Medicine*, 39(5), 487–498. Available at: http://doi.org/10.1002/ajim.1042

Arcury, T. A., Wiggins, M. F., & Quandt, S. A. (2009). Conclusions: An Agenda for Farmworker Social Justice in the Eastern United States. In S. A. Quandt, & T. A. Arcury (Eds.), *Latino farmworkers in the Eastern United States* (pp. 221–233). New York: Springer. Available at: www.springerlink.com/content/k413847031p03243/abstract/

Barndt, D. (2002). *Tangled routes: women, work, and globalization on the tomato trail.* Lanham, MD: Rowman & Littlefield.

Baum, J. (n.d.). White House Report Quantifies Farm Bill's Economic Impact. Available at: http://farmfutures.com/story-white-house-report-quantifies-farm-bills-economic-impact-0-105233 (accessed November 23, 2013).

Becker, G. (1962). Investment in Human Capital: A Theoretical Analysis. *The Journal of Political Economy,* 70(5), 9–49.

Bon Appétit Management Company Foundation and United Farm Workers. (2011). Farmworker Inventory – Bon Appétit Management Company (Inventory No. 1). Available at: www.bamco.com/uploads/documents/farmworkerinventory_0428_2011.pdf

Borjas, G. J. (1999). The economic analysis of immigration. In D. Card, & O. Ashenfelter (Eds.), *Handbook of Labor Economics* (Vol. 3, Part 1, pp. 1697–1760). Elsevier. Available at: www.sciencedirect.com/science/article/B7P5V-4FPGMDP-9/2/b760fc683ea167beca6e688c7606dfd0

Bracero History Archive. (n.d.). About Us. Available at: http://braceroarchive.org/about (accessed March 20, 2016).

Brennan, K., Economos, J., & Salerno, M. M. (2015). Farmworkers Make Their Voices Heard in the Call for Stronger Protections from Pesticides. *NEW SOLUTIONS: A Journal of Environmental and Occupational Health Policy*, 25(3), 362–376. Available at: http://doi.org/10.1177/1048291115604428

Brown, S., & Getz, C. (2011). Farmworker Food Security and Production of Hunger in California. In A. H. Alkon, & J. Agyeman (Eds.), *Cultivating food justice: race, class, and sustainability* (pp. 121–146). Cambridge, MA: MIT Press.

Callewaert, J. (2002). The Importance of Local History for Understanding and Addressing Environmental Injustice. *Local Environment*, 7(3), 257–267. Available at: http://doi.org/10.1080/1354983022000001633

Carrasquillo, N. (2011). Race and Ethnicity from the Point of View of Farm Workers in the Food System. *Race/Ethnicity: Multidisciplinary Global Contexts*, 5(1), 121–131.

Cohen, W. J., & Myers, R. J. (1950, October). Social Security Act Amendments of 1950: A Summary and Legislative History. *The Social Security Bulletin*. Available at: www.ssa.gov/history/1950amend.html (accessed March 11, 2016).

Delgado-Wise, R. (2004). Critical Dimensions of Mexico–US Migration under the Aegis of Neoliberalism and NAFTA. *Canadian Journal of Development Studies*, 25(4), 591–605.

Department of Labor. (n.d.). H-2A Certification. Available at: www.foreignlaborcert.doleta.gov/h-2a.cfm (accessed February 23, 2016).

DeWitt, L. (2010). The Decision to Exclude Agricultural and Domestic Workers from the 1935 Social Security Act. *Social Security Bulletin*, 70(4). Social Security, Office of Retirement and Disability Policy. Available at: www.ssa.gov/policy/docs/ssb/v70n4/v70n4p49.html

DREAM Act. (n.d.). Welcome to the DREAM Act Portal. Available at: http://dreamact.info/ (accessed March 27, 2016).

Durand, J., Massey, D. S., & Parrado, E. A. (1999). The New Era of Mexican Migration to the United States. *The Journal of American History*, 86(2), 518–536. Available at: http://doi.org/10.2307/2567043

EC. (n.d.). 12th Round of TTIP Talks Picks Up the Pace for a High Standard Deal – Trade – European Commission. Available at: http://trade.ec.europa.eu/doclib/press/index.cfm?id=1467 (accessed March 27, 2016).

EEOC. (n.d.a). Laws Enforced by EEOC. Available at: www.eeoc.gov/laws/statutes/index.cfm (accessed March 11, 2016).

EEOC. (n.d.b). Title VII of the Civil Rights Act of 1964. Available at: www.eeoc.gov/laws/statutes/titlevii.cfm (accessed March 11, 2016).

Farm Aid. (n.d.). How will TPP and TTIP Affect Farmers? Available at: https://www.farmaid.org/issues/industrial-agriculture/how-will-the-new-trade-deals-being-negotiated-affect-farmers/ (accessed March 27, 2016).

Farmworker Justice. (n.d.a). US Labor Law for Farmworkers. Available at: www.farmworkerjustice.org/advocacy-and-programs/us-labor-law-farmworkers (accessed March 1, 2016).

Farmworker Justice. (n.d.b). Immigration Reform & Farmworkers. Available at: www.farmworkerjustice.org/advocacy-and-programs/agjobs (accessed March 27, 2016).

Federal Register. (n.d.). Pesticides; Agricultural Worker Protection Standard Revisions. Available at: www.federalregister.gov/articles/2015/11/02/2015-25970/pesticides-agricultural-worker-protection-standard-revisions (accessed March 15, 2016).

Garcia, M. (2012). *From the jaws of victory: the triumph and tragedy of Cesar Chavez and the farm worker movement.* Berkeley, CA: University of California Press. Available at: http://site.ebrary.com/lib/alltitles/docDetail.action?docID=10595415

Gray, M. (2014). *Labor and the locavore: the making of a comprehensive food ethic* Available at: http://librarytitles.ebrary.com/Doc?id=10779332

Greenstone, E. (1975). Farmworkers in Jeopardy: OSHA, EPA, and the Pesticide Hazard. *Ecology Law Quarterly*, *5*(1), 69. Available at: http://doi.org/doi:10.15779/Z382G01

Guthman, J. (2004). *Agrarian dreams: the paradox of organic farming in California.* Berkeley, CA: University of California Press. Available at: http://site.ebrary.com/id/10064761Hahamovitch, C. (1997). *The fruits of their labor: Atlantic coast farmworkers and the making of migrant poverty, 1870–1945.* Chapel Hill, NC: University of North Carolina Press.

Hanson, G. H., Scheve, K., Slaughter, M. J., & Spilimbergo, A. (2001). Immigration and the U.S. Economy: Labor-Market Impacts, Illegal Entry, and Policy Choices. *SSRN eLibrary.* Available at: http://papers.ssrn.com/sol3/papers.cfm?abstract_id=296108

Hanson, G. H., & Spilimbergo, A. (2001). Political economy, sectoral shocks, and border enforcement. *Canadian Journal of Economics*, 34(3), 612–638.

Harris, C. (2005). *CATA policy manual: laws affecting the farmworker population and CATA's position.* New Jersey: CATA – The Farmworkers Support Committee. Available at: www.cata-farmworkers.org/english%20pages/CATA%20Policy%20Manual.pdf

Harrison, J. L. (2011). *Pesticide drift and the pursuit of environmental justice.* Cambridge, MA: MIT Press.

Haus, L. (1995). Openings in the Wall: Transnational Migrants, Labor Unions, and U.S. Immigration Policy. *International Organization*, 49(2), 285–313.

Hillman, A. L., & Weiss, A. (1999). A Theory of Permissible Illegal Immigration. *European Journal of Political Economy*, 15(4), 585–604.

Holmes, S. M. (2013). *Fresh fruit, broken bodies: migrant farmworkers in the United States.* Berkeley, CA: University of California Press.

Howell, D. (2007). Do Surges in Less-Skilled Immigration Have Important Wage Effects? A Review of the U.S. Evidence. *Working Papers: Political Economy Research Institute*, 128, 1–18.

IFOAM (International Federation of Organic Agriculture Movements). (n.d.a). Leading, Uniting and Assisting the Organic Movement since 1972. Available at: www.ifoam.org

IFOAM (International Federation of Organic Agriculture Movements). (n.d.b) The Full Diversity of Organic Agriculture: What We Call Organic. Available at: http://infohub.ifoam.org/sites/default/files/page/files/full-diversity-organic-agriculture-en.pdf

International Trade Administration. (n.d.). Available at: http://trade.gov/index.asp (accessed February 15, 2016).

James, S., & Griswald, D. (2012, December 14). Freeing the Farm: A Farm Bill for All Americans. Available at: www.cato.org/publications/trade-policy-analysis/freeing-farm-farm-bill-all-americans (accessed November 23, 2013).

Kandilov, A. M. G., & Kandilov, I. T. (2010). Job Displacement from Agriculture. *American Journal of Agricultural Economics*, 92(3), 591–607. Available at: http://doi.org/10.1093/ajae/aaq021

Kominers, S. (2015). *Working in fear: sexual violence against women farmworkers in the United States.* Oxfam America, Inc. Available at: www.northeastern.edu/law/pdfs/academics/phrge/kominers-report.pdf

Martinez, R. L. (1993). NAFTA's Effect on Human Rights at the Border. *U.C. Davis Law Review*, 27, 979.

Marx, J. (2016, February 11). Trans-Pacific Partnership: A Guide to the Deal. *The Desert Sun*. Available at: www.desertsun.com/story/news/2016/02/11/tpp-trans-pacific-partnership-obama-sunnylands-explainer/80189584/

Massey, D. S. (1988). Economic Development and International Migration in Comparative Perspective. *Population & Development Review*, 14(3), 383–413. Available at: http://doi.org/Article

Massey, D. S., & Espinosa, K. E. (1997). What's Driving Mexico-U.S. Migration? A Theoretical, Empirical, and Policy Analysis. *American Journal of Sociology*, 102(4), 939. Available at: http://doi.org/Article

Meyers, E. (2000). Theories of International Immigration Policy: A Comparative Analysis. *International Migration Review*, 34(4), 1245. Available at: http://doi.org/10.2307/2675981

Mitchell, D. (2012). *They saved the crops: labor, landscape, and the struggle over industrial farming in Bracero-era California*. Athens, GA: University of Georgia Press.

Morison, J., Hine, R., & Pretty, J. (2005). Survey and Analysis of Labour on Organic Farms in the UK and Republic of Ireland. *International Journal of Agricultural Sustainability*, 3(1), 24–43. Available at: http://doi.org/10.1080/14735903.2005.9684742

NASDA. (n.d.). NASDA's Comments on EPA's Proposed Certification of Pesticide Applicators Changes. Available at: www.nasda.org/Policy/filings/10937/40348.aspx (accessed March 20, 2016).

National Farm Worker Ministry. (n.d.). Labor Laws. Available at: http://nfwm.org/education-center/farm-worker-issues/labor-laws/ (accessed March 1, 2016).

NSAC. (n.d.). What is in the 2014 Farm Bill for Sustainable Farms and Food Systems? Available at: http://sustainableagriculture.net/blog/2014-farm-bill-outcomes

Obach, B. K. (2004). *Labor and the environmental movement: the quest for common ground.* Cambridge, MA: MIT Press. Available at: http://search.ebscohost.com/login.aspx?direct=true&scope=site&db=nlebk&db=nlabk&AN=122537

OSHA. (n.d.a). Content Document. Available at:www.osha.gov/pls/oshaweb/owadisp.show_document?p_table=OSHACT&p_id=2743&p_text_version=FALSE (accessed March 13, 2016).

OSHA. (n.d.b). Safety and Health Topics: Agricultural Operations – Hazards & Controls. Available at: www.osha.gov/dsg/topics/agriculturaloperations/hazards_controls.html (accessed March 13, 2016).

OSHA. (n.d.c). OSHA Safety and Health Topics. Agricultural Operations. Available at: www.osha.gov/dsg/topics/agriculturaloperations/ (accessed March 13, 2016).

OSHA. (n.d.d). Field Sanitation. – 1928.110. Available at: www.osha.gov/pls/oshaweb/owadisp.show_document?p_table=STANDARDS&p_id=10959 (accessed February 23, 2016).

Roberts, K. D. (1997). China's "Tidal Wave" of Migrant Labor: What Can We Learn from Mexican Undocumented Migration to the United States? *International Migration Review*, 31(2), 249–293.

Rodman, S., Barry, C., Clayton, M., Frattaroli, S., Neff, R., & Rutkow, L. (2016). Agricultural Exceptionalism at the State Level: Characterization of Wage and Hour Laws for U.S. Farmworkers. *Journal of Agriculture, Food Systems, and Community Development*, 89–110. Available at: http://doi.org/10.5304/jafscd.2016.062.013

Rothenberg, D. (1998). *With these hands: the hidden world of migrant farmworkers today*. New York: Harcourt Brace & Co.

Rytina, N. (2002). *IRCA legalization effects: lawful permanent residence and naturalization through 2001*. The Cloister, Mary Woodward Lasker Center, NIH Main Campus, Bethesda, MD: Office of Policy and Planning, Statistics Division, U.S. Immigration and Naturalization Service. Available at: www.dhs.gov/xlibrary/assets/statistics/publications/irca0114int.pdfSchlosberg, D. (2004). Reconceiving Environmental Justice: Global Movements and Political Theories. *Environmental Politics*, 13(3), 517–540. Available at: http://doi.org/10.1080/0964401042000229025

Schlosberg, D. (2007). *Defining environmental justice: theories, movements, and nature*. New York: Oxford University Press.

Schultz, D. (2016). 11th Hour Change to EPA Farm Rule Angers Capitol Hill. Available at: www.bna.com/11th-hour-change-n57982074301/ (accessed 23 June 2016).

Scott, R. E. (2003). *The high price of "free" trade: NAFTA's failure has cost the United States jobs across the nation* (pp. 1–13). Washington, DC: Economic Policy Institute. Available at: http://docs.google.com/viewer?url=http://www.epi.org/page/-/old/briefingpapers/147/epi_bp147.pdf&hl=en_US&embedded=true

Sen, A. (2009). *The idea of justice*. Cambridge, MA: Belknap Press of Harvard University Press.

Shiva, V. (2005). *Earth democracy: justice, sustainability, and peace*. Cambridge, MA: South End Press.

Shreck, A., Getz, C., & Feenstra, G. (2006). Social Sustainability, Farm Labor, And Organic Agriculture: Findings from an Exploratory Analysis. *Agriculture and Human Values*, 23(4), 439–449.

Starmer, E. (n.d.) Helping Organics Grow with Clear Livestock and Poultry Standards. USDA Blog. Available at: http://blogs.usda.gov/2016/04/07/helping-organics-grow-with-clear-livestock-and-poultry-standards/

USDA (United States Department of Agriculture). (2012, December 5). Agricultural Marketing Service: Organic Labeling. Available at: www.ams.usda.gov/AMSv1.0/NOPOrganicLabeling

USDA (United States Department of Agriculture). (2016). Blog, n.p., Web. 8 June.

U.S. Department of Labor. (2008, July). *Fact Sheet #49: The Migrant and Seasonal Agricultural Worker Protection Act. Wage and Hour Division*. Available at: www.dol.gov/whd/regs/compliance/whdfs49.pdf

United States Farmworker. (n.d.). Fact Sheet: Coalition of Immokalee Workers. Available at: www.ciw-online.org/fwfactsheet.pdf

United States Trade Representative. (n.d.) Transatlantic Trade and Investment Partnership (T-TIP). Available at: https://ustr.gov/ttip (accessed March 27, 2016).

Vogl, C., Kilchner, L., & Schmidt, H. (2005). Are Standards and Regulations of Organic Farming Moving Away from Small Farmers' Knowledge? *Journal of Sustainable Agriculture*, 26(1), 5–26.

Wiggins, M. (2009). Farm Labor and the Struggle for Justice in the Eastern United States. In T. Arcury, & S. Quandt (Eds.), *Latino farmworkers in the Eastern United States* (pp. 201–220). New York: Springer.

Wilson, K. L., & Portes, A. (1980). Immigrant Enclaves: An Analysis of the Labor Market Experiences of Cubans in Miami. *American Journal of Sociology*, 86(2), 295–319. Available at: http://doi.org/Article

4 Case study in conventional agriculture

The Farmworker Association of Florida

Introduction to the Farmworker Association of Florida

I first encountered the Farmworker Association of Florida (FWAF) while I was living in Orlando and working in higher education, first at a small, private liberal arts institution in a position focused on issues related to social change, and next as an instructor for a Business Ethics and Social Responsibility course at a different institution. Through participation in their organization with programs like the Toxic Tour, the Bandana Project, and the Lake Apopka Farmworker Memorial Quilt Project, in addition to engaging my students in pesticide health and safety trainings, pursuing dialogue on the ramifications of current capitalist business models and their impact on workers, and traveling to several nurseries and farms to work alongside the farmworkers, the stories of the people that this organization represents gripped me as they have many others. Over the course of my eight-plus-year relationship with FWAF, I never cease to be in awe of the sheer scale of the work they do, and their continual motivation to lift up farm work as a profession that is valued with dignity for all who have suffered in farm labor historically.

In this chapter I detail their work, focusing on the work of the Apopka area office, with which I worked most closely. The story of this area, its farmworkers, and the work of the organization exemplifies and magnifies many of the issues discussed and detailed in Chapters 2 and 3, specifically as they pertain to workers in large-scale, corporate, conventional agriculture. South Apopka has been the subject of many articles, books, and even short films, and after spending time there it is easy to see why—it's hard to believe that what has happened, and continues happening there, is real and that most people are unaware of the realities that face the people who harvest our food. It is as Nolan Kline and Rachel Newcomb assert:

> Our vision of American farm life was shattered the day we first drove across the street from the Farmworker Association office into a world of intense poverty and suffering. Although the journey was not far in physical distance, it took us far away from the romanticized, almost American Gothic perceptions of farming that most Americans have and

> plunged us into the frightening realities of farm workers living in shacks with dirt floors.
>
> (2013, p. 169)

In addition, the work of the organization provides a helpful bridge to justice claims and the question "How should things be?", which will be the focus of Chapter 6, through its active work involving and recognizing the farmworker community in all it does.

The Farmworker Association of Florida was founded informally in 1983 when farmworkers in Central Florida came together as a group to address the problems of the farmworker community. In doing so, they sought to organize themselves more effectively in their struggle for better housing, wages, and working conditions. In 1986, they incorporated as the Farmworker Association of Central Florida (later the name changed to the Farmworker Association of Florida) and expanded into a statewide organization in 1992, concentrated in Central and South Florida. Years later, workers in the community found solace with one another as people they had worked the fields with for decades began to die from diseases such as emphysema and lupus, many afflicted at disproportionately young ages (Berkey, 2012; La Asociación Campesina de Florida/Farmworker Association of Florida, n.d.c). Beyond finding those with similar experiences with whom to commiserate, the farmworkers began to explore their own force for change in the very systems that were negatively impacting their lives.

While it has grown over the years, the Farmworker Association remains a grassroots, community-based organization of over 10,000 member families who work primarily in the vegetable, citrus, mushroom, sod, fern, and foliage industries in 15 counties throughout Central and South Florida. The members are approximately 94 percent Latino (predominately Mexican, Guatemalan, and Salvadoran), 3 percent Haitian, and 3 percent African American, and approximately 40 percent are women. In total, the Farmworker Association of Florida's organizing efforts affect at least 40,000 individuals in rural agricultural areas each year. Each of the populations that comprises its membership has its own unique set of needs and issues, ranging from the multiple issues that arise with immigrant and migrant laborers to the long-term health issues of those who have worked farms in the community for generations (Kamel et al., 2001; Kline, 2010; La Asociación Campesina de Florida/Farmworker Association of Florida, n.d.b).

Staying true to its roots as an organization founded by farmworkers, the composition of the Board of Directors is reflective of the association's commitment to be led and governed by the workers themselves. It is responsible for setting general policy, guaranteeing the implementation of work plans, and monitoring Association finances. The organization receives funding through private foundation grants. In addition, many of the programs receive state and federal grant funding. The Board consists largely of current or former farmworkers, and members are elected by local membership councils using a formula that reflects the ethnic/racial/gender diversity of the farmworker community it represents. Through its Board makeup, the Farmworker Association of Florida seeks to identify its community assets and show that it is community-driven and controlled, and as the Chair of the Board puts it:

> We are committed to the success of the Farmworker Association because we believe in the organization. It is our organization, and we are proud to have a successful group that listens to the workers, and involves them in addressing their problems and improving their communities.
>
> (Central Florida Foundation, 2014, p. 4)

Apopka, Florida, is a suburb of Orlando that has farmworker populations who work in a variety of crops, including but not limited to cabbage, corn, cucumbers, citrus, ornamental plants, and ferns. This area is home to one of the largest lakes in Florida, Lake Apopka, and therefore historically was home to a number of muck farms. Muck farming techniques involve flooding the lands during the hot summer months to keep them moist and fertile and to inhibit weed growth, and then draining the lake back down during the growing months to use the land for crop production. This practice occurred regularly in this area from the 1940s to the 1990s without consideration of its impacts, particularly in relation to the pesticides and chemical fertilizers being used on the farm land entering the body of water as it was flooded and drained over the years. According to a briefing prepared by the Florida Senate (2011, p. 1):

> The decline of Lake Apopka can be traced to the loss of 20,000 acres of wetlands along the lake's north shore to farming operations in the 1940s, agricultural discharges laden with phosphorus until the late 1990s, treated wastewater discharges from shoreline communities prior to the 1980s, discharges from citrus processing plants prior to the 1980s, organic compounds from nearby fiberglass and plastics manufacturing companies, and other industries located within the community, two local industrial landfills, two Superfund sites on Lake Apopka, and a medical waste incinerator.

In 1995, a study was published by Louis Guillette, Timothy Gross, Denise Gross, Andrew Rooney, and H. Franklin Percival that detailed reproductive abnormalities of alligators in Lake Apopka due to the pesticides in the water and their impacts on reproductive, endocrine, and immune systems in animals (Guillette et al., 1995). The following year, the Legislature passed the Lake Apopka Improvement and Management Act aimed at saving the lake by purchasing the muck farms surrounding it to cut off the source of the pollution (Florida Senate, 2011).

In the years that followed, the Legislature developed guidelines for those buyouts and the proceeds of those buyouts, little of which went to training or otherwise assisting the labor force. As such, upon the farms closing, around 2,500 (with some estimates as high as 5,000) farmworkers were left without work, with around only 20 percent of the funds captured from selling the farmland and farm equipment going toward their retraining and re-employment. According to the brief by the Florida Senate, "Farm workers, who earned a living from these farming operations for years and even generations, lost their livelihoods when the state purchased the farms from the owners and some, who lived in company-provided housing, no longer had that option" (ibid., p. 6). As Barry Estabrook (2012, p. 32) states:

> The 2,500 workers, who often had families who lived with them on the land, got nothing other than orders to clear out of the shacks and trailers that belonged to the landowners. Proposals to fund retraining programs to prepare them for other jobs during the final years of agriculture were opposed by the landowners, who did not want a single laborer to leave until the last carrot was pulled, the last bean picked.

Not only were the former farmworkers' livelihoods threatened in these ways, but their health was at stake as well. As stated in the indictment from the Permanent People's Tribunal on Agrochemical Transnational Corporations:

> The African-American Lake Apopka farm worker community is an ageing community that in many cases had parents, grandparents, children and grandchildren working alongside each other in the fields. Hence, the bioaccumulative properties of these pesticides have negatively impacted multiple generations of farm workers with long-term, chronic health problems from cumulative and synergistic effects.
>
> (Permanent People's Tribunal, 2011, p. 194)

This same tribunal cited violations of the right to health, the right to a healthy environment, and the right to livelihood, among those for which the chemical companies that produced the pesticides were responsible.

In 2011, after years of work to seek accountability for the health harms to farmworkers due to their chronic exposure to persistent organic pollutant pesticides, such as DDT, Toxaphene, and others, the local community of Lake Apopka farmworkers was able to celebrate a victory of sorts when the international Permanent Peoples Tribunal (PPT) issued a verdict of guilty in a case against the "Big Six" (BASF, Bayer, Dupont, Dow, Monsanto, and Syngenta) agrochemical companies for violation of human rights to health and a healthy environment due to exposure to toxic pesticides (The Permanent People's Tribunal, n.d.). Pesticide Action Network (PAN) International built the legal case against the "Big Six" by identifying communities and peoples around the globe whose lives, environments, and livelihoods had been adversely impacted by the pesticide products these companies produced. Working closely with its ally PAN, the FWAF Lake Apopka farmworker community was one of the few communities chosen from North America to be included in the legal case before the PPT. While the victory was mostly symbolic, since the PPT has no international legal enforcement authority, it was a validation for the community's concerns and the case did garner media attention throughout India and other Asian countries. In addition, it did send a strong message to the world and to the multinational pesticide corporations of the harmful effects of the widespread use of these toxic chemicals.

During the time when the farms were shutting down, health considerations of the farmworkers were unacknowledged, the farmworkers who had been exposed regularly and often over the course of many years to the same chemicals that had caused such adverse and long-lasting reactions in the alligators of the lake. Further, in 1998, around 1,000 fish-eating birds were found dead in this area, adding further ecological evidence to the harmful health impacts of exposure to pesticides and other environmental contaminants in the lake due to their use on the farmland surrounding it (Florida Senate, 2011). Community members voiced concern over the health impacts of their exposures. As Dale Slongwhite put it in *Fed Up: The High Costs of Cheap Food*:

> Once the EPA had linked the bird deaths to high levels of pesticide exposure, the farmworkers began to clearly articulate what they wanted: an official health study to determine whether the chemicals found in the birds were present in significant amounts in their own bodies. If so, they wanted to know what that would mean for their health.
>
> (2014, p. 11)

In 2005, Ron Habin, a local anthropologist, and Geraldean Matthew, a long-time farmworker leader of the Farmworker Association of Florida, worked along with other community leaders to conduct a health survey through interviews that allowed farmworkers to assess their health problems along with their exposure to pesticides and other contaminants (Habin & Matthew, 2006). Some 92 percent of their participants indicated that they were exposed to pesticides in the workplace; 83 percent stated that they were in "fair" or "poor" health; 85 percent feel their exposure to pesticides has affected their health; and 79 percent believe that their pesticide exposure is directly related to current health problems. According to Nolan Kline (2010, p. 127), "The farmworkers of Apopka have a myriad of serious health concerns, including lupus, high rates of cancer, diabetes, hypertension, congestive heart failure, and high rates of birth defects among children and grandchildren."

In addition to their work with the African-American community in Apopka, all offices of the Farmworker Association of Florida work on immigrants' rights, farmworker health and safety, worker rights, and civic participation, particularly given the large numbers of immigrant farmworkers in Florida. In her book, Dale Slongwhite interviews Tirso Moreno, one of the founders of the FWAF. He discusses the changing demographic of farmworkers in Florida and how their various branches evolved, based on both disaster response and the expressed needs of the farmworker community. In the 1960s and 1970s, Moreno explains that the majority of farmworkers in the area "were black – either African

American from other states in the South or African descendants from the Caribbean" (Slongwhite, 2014, p. 138). Over time, this demographic has changed, as have FWAF's focus on and capacity for supporting immigrant farmworkers on issues that pertain to them, including access to health and safety-related information; adequate healthcare; information regarding their rights in relation to fair wages, housing, and working conditions and potential paths to citizenship; as well as organizing for immigrants' rights locally and nationally.

The Farmworker Association of Florida has had many major accomplishments in its 30-year history as an organization. It categorizes these accomplishments in four ways: (1) legal accomplishments; (2) community-mobilizing accomplishments; (3) education and research accomplishments; and (4) collaborative accomplishments (La Asociación Campesina de Florida/Farmworker Association of Florida. n.d.c). Its legal accomplishments include the successful passage of the Alfredo Bahena Act, or the Farmworkers' Right-to-Know Law, in 2004 and the Farmworker Transportation Safety Act in 2006, along with advocacy and a lawsuit that resulted in the coverage of ferncutters and fernery workers under the Agricultural Worker Protection Act.

Its community-mobilizing accomplishments occur on local and regional levels, and have included organizing workers and community members to secure improvements in wages and working conditions for farmworkers in over 60 Central Florida companies. It was also integral in the organization of the first tri-lingual International Farmworker Forum in 1998 and in organizing the immigrant community for the May 1st, 2006 Immigrants' Rights Rally and March in downtown Orlando, which attracted 30,000 participants. Politically, the Farmworker Association of Florida members assisted community members in voter research, voter registration, and mobilization to vote in elections in order to shift the composition of the local government to be more inclusive and reflective of the low-income community in the farmworker community of Fellsmere, Florida. It also created the Lake Apopka Dislocated Workers Program, which outreached to over 1,200 farmworkers on issues such as job retraining and reemployment opportunities, as well as housing relocation.

One of its first most notable accomplishments was its status as a founding member, along with several other grassroots farmworker groups, of the Farmworker Health and Safety Institute's extensive educational program. Among FHSI's early accomplishments were the development of an EPA-certified pesticide safety training based on popular education methodology, which informed workers of their rights to a safe workplace and how to protect themselves against pesticide exposure (they have trained more than 7,000 farmworkers); and the development of the "diagnostico," a workplace assessment tool to identify and document employer violations of worker protections, and other workplace hazards. Since the early days of training farmworkers, FWAF's commitment to education to improve farmworker health has steadily grown, and now includes curricula to train farmworkers on the impact of farm work on reproductive health, and curricula to train healthcare providers on the detection, diagnosis, treatment, and reporting of pesticide exposure in farmworkers.

The Association facilitated the administration of the aforementioned community-generated Lake Apopka Farmworkers Environmental Health Survey project, with community members interviewing 148 former farmworkers impacted by pesticide exposure. The Association oversaw the completion of more than 600 diagnostics for monitoring violations of worker protection standards and field sanitation laws. It has assisted in the creation and development of successful economic development projects, such as PEP Labor Crews citrus harvesting cooperative and the La Tienda ethnic food stores. They also have several ongoing community-based participatory research partnerships with universities that center on farmworker health (La Asociación Campesina de Florida/Farmworker Association of Florida, n.d.c). While the organization no longer focuses on this particular program, its successes highlight the involvement of the FWAF and its former work while providing insight into its evolutionary changes as an organization.

Over the decades, the Farmworker Association of Florida has co-founded a number of coalitions and collaborations to advance their work of justice for farmworkers. These included the Farmworker Health and Safety Institute (FHSI, discussed above), the Farmworker Network for Economic and Environmental Justice, the Florida SAFER Coalition, the Latino Immigrant Network of the Southeast, the Florida Immigrant Coalition and others, some of which continue to be active today. Currently, FWAF is an active member of the Rural Coalition, the Food Chain Workers Alliance, the Domestic Fair Trade Association, and the U.S. Food Sovereignty Alliance, among others. FWAF actively participated in the National People of Color and Indigenous People Environmental Leadership Summits, the United Nations World Conference against Racism, two World Social Forums, the U.S. Social Forum, and the Southeast Social Forum. In addition, the Farmworker Association of Florida is one of only four U.S. members of La Via Campesina (LVC), an international movement of peasants, small and medium-sized producers, landless farmers, rural women, indigenous people, rural youth, and agricultural workers. LVC is committed to defending the values and the basic interests of its members, and is an autonomous, pluralist and multicultural movement, independent of any political, economic, or other type of affiliation. Through this membership, international exchanges influence and shape the work of the FWAF with a global perspective of solidarity and change. The FWAF is also part of the Farmworker Health and Justice Work Group or Team of the Coming Clean Coalition, which is a national coalition of organizations working together on environmental justice and toxics issues. The FHJ Work Group works together to promote farmworker health and justice (Farmworker Health and Justice, n.d.). The other member organizations of the Team include California Rural Legal Assistance Foundation, Earthjustice, Farmworker Justice, Migrant Clinicians Network, Pesticide Action Network North America, and Toxic Free North Carolina. Some of these organizations and their work will be detailed in Chapter 6.

FWAF organizes its work through a number of different programs, detailed on its website (La Asociación Campesina de Florida/Farmworker Association of Florida. n.d.b). These include programs on worker justice, pesticide actions, immigrants' rights, health and safety education, disaster response, civic participation, Lake Apopka environmental justice, vocational rehabilitation outreach, reproductive health, and, most recently, on agroecology, including the establishment of local community gardens in three FWAF communities. Each of these project areas has multiple initiatives, ranging from pesticide policy advocacy to voter registration, and leadership development training, among others. In addition, FWAF's work has included an increasing emphasis on community-based participatory research in partnership with academic institutions around issues of occupational risks to farmworker health. This has included, more recently, a study with Emory University on the reproductive health of Florida farmworkers and an ongoing study on heat stress among farmworkers in Florida.

One example of a project related to worker justice was a USDA-funded Sustainable Agriculture Research and Education Project, in which FWAF worked in partnership with the Florida Organic Growers Association to expand organic and sustainable food system models to the southern states. The model they used for this project is the Agricultural Justice Project, which focuses on and develops standards for a food system that is both economically viable, while combining dimensions of pesticide-free agriculture, social equity and just working conditions for farmworkers.

A more recent initiative of the organization is the agroecology project, which was launched in 2015 with the Campesino-a-Campesino Agroecology Encuentro. Originally a community-based initiative that began in 2010, the program provides space for dialogue and cooperative work surrounding food sovereignty, farmworker struggles, human and environmental health, and transnational agricultural justice. Agroecology as a means to share knowledge and promote the value of chemical-free farming, intergenerational learning, traditional and ethnic foods and farming methods, and cooperative practices builds, strengthens, and empowers local communities, while joining in solidarity in the movement with the struggles of small farmworkers and peasant farmers around the world. Workshops on the traditional wisdom of agroecology and the values of respect in relation to food and the earth are structured to share regionally and technically appropriate methods that emphasize community involvement, agency, and participation (Farmworker Association of Florida, 2015). With the farmworker communities around the world ever changing and varying by region, objectives, and experiences, the overlapping and distinct goals within the agroecology project address key distinctions and connections that aid in articulating the objectives of each farmworker community.

Lastly, related to its work around worker justice, FWAF includes its project on the Latino Farmers Outreach Initiative (La Asociación Campesina de Florida/Farmworker Association of Florida, n.d.e), the aim of which is to assist Latinos, including current or former farmworkers, to start their own agricultural operations and small agricultural business ventures. All of these projects focus on justice for and with workers themselves, including the quilt project, a major initiative of the FWAF discussed later in this chapter.

Situating the work in the continuum of farm labor and environmental justice claims

In what follows, I use the model set forth in the Introduction to explore how the work done by the Farmworker Association of Florida is illustrative of the environmental justice concerns faced by workers in large-scale, conventional agriculture. Building on the background information provided about FWAF, I will detail the process claims/social dimensions of the plight of the farmworkers they serve, the evidence claims/environmental and structural dimensions of the injustice they face, and the justice claims/how the work of FWAF represents calls for recognition, participation, and capabilities within environmental justice.

Process claims (social dimensions)

Who?

As previously discussed, while the organization as a whole works statewide with African American, Haitian, and Hispanic farmworkers, the Apopka office of FWAF represents two main populations or groups of farmworkers: African American former farmworkers who had farmed the land for generations, and immigrants who work on the fields and nurseries in Florida for all or part of the year, some of whom migrate with the seasons. In many ways, the workers they represent and who comprise their staff, board, and membership are reflective of the larger landscape of farmworkers in the United States as discussed in Chapter 2, workers from marginalized and historically exploited groups. Clearly these groups have overlapping concerns when it comes to things like housing, wages, health, and safety, but many times they experience these things in different ways.

Stories told by the African American farmworkers and former farmworkers bring one to tears, as they detail not only their own exposure to the very pesticides that shut down many of the farms surrounding Lake Apopka, but talk of how these same chemicals have influenced their families over time, including their children and grandchildren.[1] I will never forget one of the first times I visited FWAF and met Geraldean Matthew. She talked about working in the fields as a young girl in the hot Florida sun and feeling relief at a sudden rain, only to realize it was a plane flying overhead coating the field she was working in with pesticides. Her story is recounted in Slongwhite, and she says:

> As far back as 1974, we talk among ourselves because we knew that when we go home at night there was something going on with our bodies. When we come to work the next morning, we'd tell each other about the

> toothache we had that was so bad, or how our eyes was burnin' so bad, that our skin was burnin' so bad. We talk about our headaches or that we was coughin' all night long. In the fields, if you go to talkin' about you got sick because of the pesticides, there was a hush mouth, because if you didn't keep your mouth closed, they would retaliate against you … If you go right today to a farmworker's house that worked twenty to thirty years ago, they got a spit bucket by their bed. All night long till now, they spit all night. And it's no cold.
>
> (2014, p. 39)

Geraldean has worked tirelessly for years not only to tell her story, but to pave the way for others to tell theirs. In addition, she has been instrumental in the fight for these abandoned workers to receive healthcare and resources for the abuses they suffered and of which they continue to bear the burden. She is a representative of the plight of this group of workers, and continues, like many of them, through poor and declining personal health, to fight for justice.

FWAF also serves immigrant farmworkers, who make up a substantial proportion of farmworkers nationwide. In addition to the ways that their multiple campaigns intersect with justice issues facing all farmworkers, including immigrants, their website states, "Low-income immigrant communities have many factors or barriers that contribute to them being marginalized, underserved, and underrepresented. FWAF works to inform immigrant farmworkers of their rights and engage them in decision-making and policy-setting processes that impact their lives" (La Asociación Campesina de Florida/Farmworker Association of Florida, n.d.a). Not only has the organization been active in policy advocacy impacting all immigrants and immigrant farmworkers, such as the DREAM Act and AgJOBS, but it also works with other organizations nationwide to ensure that immigrant farmworkers' voices are represented in the protections that are in place, including advocating for accessibility and language surrounding pesticide labeling and training (Farmworker Health and Justice Team, n.d.). As stated on the website for Protect Farmworkers, with which FWAF is involved:

> The people who feed us deserve no less. For too long, we as a nation have depended on a food supply that has put millions of people, including children, in harm's way of direct and indirect exposure to agricultural chemicals in the fields and greenhouses where they work and near to where they live.
>
> (Protect Farmworkers, n.d.)

As described in Chapter 2, the justice issues impacting many immigrant farmworkers are exacerbated due to their lack of legal immigration status in the United States, and other barriers to reporting workplace violations, such as lack of knowledge of the regulations and their rights as workers, intimidation tactics by contractors, supervisors and/or employers, language barriers, fear of retaliation and job loss, lack of a support network, residing in remote, rural areas with no access to transportation and/or assistance, among others. Kamel et al. name this challenge when it comes to studying health impacts on immigrant and migrant populations, explaining:

> One reason for the scarcity of such studies is that researchers perceive the farmworker population to be relatively inaccessible. Farmworkers may move frequently, making it difficult to define and enumerate a target population. Frequent movement, as well as immigrant status, lack of formal education, language barriers, and isolation from the larger community may decrease response. In addition, the nature of agricultural work makes it difficult for occupational health researchers to gain access to the workplace.
>
> (2001, p. 457)

They offer some advice in their study about how to overcome these obstacles to gain information through use of community members and recruiters who have knowledge of the farmworker population and who can or may be willing to participate in studies. Kline describes the success of the FWAF in providing services to the immigrant population despite these challenges, but cites less success when providing similar programs and services to former farmworkers, stating that:

> This lack of success is largely because of structural hindrances and the different scope of needs for each population. The FWAF can provide services for current farmworkers, such as pesticide and safety training, but they are unable to meet the more challenging needs of the former farmworker population, whose deteriorating health requires assistance from medical specialists.
>
> (2010, p. 127)

In essence, the work of the Association with immigrant farmworkers is proactive work that is happening synchronously with their work in the fields, meaning that as policies change and training, education, and resources become available, they are able to share those with the immigrant populations. There is little being done to make up for the lack of those policies and protections that impact the former farmworker community to this day.

Why is this happening and how did it come to be?

Earlier in this chapter, I detailed the brief history of the Lake Apopka area that, in addition to information offered throughout this book, provides the background of justice issues faced by the two main populations served by the Farmworker Association of Florida: immigrant and non-immigrant farmworkers. At the request of the FWAF, Laura Bermudez (2012) conducted a study that focused on two areas: first, documenting cases of worker abuse and pesticide exposure, and second, to note the discrepancies between what was studied and the number of cases reported to the State of Florida as represented in official, public statistics. Guiding questions for her research were, "Why aren't real cases of abuse and pesticide exposure being reflected in the official statistics? Why do farmworkers remain silent about what they experience in their jobs?" (ibid., p. 3). In conducting the study, and as implied in its title *Farmworkers in Florida: Silence Is Their Rational Choice*, Bermudez highlights the central tenet to the environmental justice process claims facing both groups of workers—that they do not feel empowered, entitled, or safe to speak up and report violations in the workplace. In the case of former farmworkers, the detail throughout all of the research done on the Lake Apopka workers was the fact that they knew that something was wrong but were told to keep quiet about it. As it pertains to current farmworkers who are predominantly immigrant, they feel unsafe due to citizenship status and anti-immigrant sentiment, language barriers, poverty, and threats of job loss and/or intimidation and harassment, or are simply unaware that they have worker rights and that there are formal channels through which they can report. Bermudez adds to this, outlining her findings on workers' decision to stay silent as: few work alternatives, dependants back home, children's health issues, their own health issues, working while pregnant, and living on the property (ibid., pp. 10–15).

Evidence claims (environmental/structural dimensions)

What?

As I have begun to explore, the evidence claims, or environmental and structural dimensions of injustice these workers and former farmworkers face are extensive. In her study, Bermudez (2012) distills her findings to the following areas of worker abuse and pesticide exposure: issues with sanitation, being pushed to work rapidly in a dangerous environment, lack of safety equipment, covering the costs of work-related injuries, no overtime, pesticide application practices, cases of pesticide exposure, and long-term and intergenerational effects of pesticide exposure (ibid., pp. 4–10). Her findings mirror those issues detailed in Chapter 2, and play out in myriad ways in the Central Florida farmworker community.

Occupational hazards and pesticide exposure

There is regular media coverage of the impacts of occupational hazard, particularly pesticide exposure, in the Apopka area farmworker community. One such article discusses Geraldean Matthew, stating:

> In 1988, Matthew was diagnosed with lupus, and in 2007 she was told by doctors that she had chronic kidney disease. Matthew also has diabetes and liver failure. Today, two of her adult daughters have lupus. Another adult

> daughter suffered a stroke at 3 years old that left her left side partially paralyzed. Lupus is a chronic inflammatory disease that occurs when the body's immune system attacks organs and tissues. Lupus can affect different body systems, including the kidneys, blood cells, brain, heart and lungs
>
> (Comas, 2015)

The same article decries the lack of research on the farmworkers themselves, noting that:

> It's not clear how many former farmworkers are sick with lupus or chronic kidney disease today… a 2002 investigation by the *Orlando Sentinel* found a variety of illnesses in 50 people tied to the muck farms, including 20 cases of lupus. Other illnesses included cancers, rheumatoid arthritis and respiratory problems.
>
> (ibid.)

Barry Estabrook, interviewed in Sanjay Rawal's short film on Lake Apopka (Earthjustice & Rawal, 2015), notes that exposure to pesticides as the cause of the rampant nature of these diseases among former farmworkers in this area is nearly impossible to prove.

Because of this, former farmworkers are caught in a difficult conundrum, as explained by Christopher Balogh in his piece for the *Orlando Weekly* (2011):

> Without evidence that they've been hurt by pesticides, no one will study their health problems; but without anyone studying their health problems, they can't show evidence that pesticides are what has hurt them. Although medical professionals suspect a higher likelihood of lupus occurring in a person who is in direct contact with pesticides, and studies show that pesticides could have some relationship to the development of lupus, there is no conclusive evidence that supports it.

But not all is lost, and there are some promising models for how farmworkers are involved and have been involved in studying and better understanding their own health and the causes of illness. In their article "Farmworkers and Pesticides: Community Based Research," Thomas Arcury, Sara Quandt, and Linda McCauley (Arcury et al., 2000) discussed a workshop they conducted to bring together those working and/or conducting research with farmworker communities. While they did not solve the problematic cycle described above, they offer concrete solutions for conducting research. They identify two areas that need further development to better understand and study the impacts of pesticide exposure: (1) rigorous epidemiological and survey research that produces generalizable results; and (2) the need for procedures to measure biological exposure to pesticides among farmworkers (ibid., p. 791). Advances in these areas have occurred, and while obstacles still remain, some of the collaborative research being done by the FWAF is leading the way.

Another area that FWAF explored was in examining and promoting ways that employers can educate their workers on and better monitor pesticide safety behavior. According to a journal article by Brian Mayer, Joan Flocks, and Paul Monaghan (Mayer et al., 2010), FWAF was a partner in a study that looked at the role of the attitudes and behaviors of employers and supervisors in making pesticide safety a central concern. While their study did unveil some troubling statistics about pesticide exposure and its subsequent health impacts, it also revealed that hand washing has a huge impact on health and exposure, and that hand washing can be positively influenced by the workplace. In addition to this and the activism discussed earlier in the chapter, the FWAF also utilizes education and community organizing and, "works to improve farmworkers' health, working conditions, and access to quality health care; raise awareness of the harmful effects of toxic pesticides; and influence policy related to health and safety protections for farmworkers" (La Asociación Campesina de Florida/Farmworker Association of Florida, n.d.d). This study has since concluded, partially due to employer resistance.

Housing and wage theft

In collaboration with the FWAF, researchers Joan Flocks and Alan Burns (2006) conducted focus group and interview research on the housing-related issues facing farmworkers. The themes that emerged from their data regarding housing included lack of affordability in relation to income, problems of availability of and access to both private and subsidized housing, the location of housing and its proximity to the farm and/or nearby communities, and the home environment, both its safety and security. Given that they talked with both farmworkers and farmworker stakeholder groups, including employers, academics, and advocates, they found that:

> In general, stakeholders agreed that workers want and should have housing that is clean and sturdy. Beyond that, there were more varying opinions about issues such as location and the types of housing that best serve a particular demographic group. For example, many stakeholders felt that farmworker housing should be located within communities and not isolated from them.
>
> (ibid., p. 64)

While these issues are not unique to the FWAF farmworker groups and are representative of the housing-related inadequacies facing all farmworkers, they highlight how these workers face issues that are illustrative of and reflected in those faced by farmworkers more broadly. To address these issues, the authors suggest three things: (1) to promote home ownership; (2) to improve the enforcement of health and housing regulations; and (3) to pay farmworkers higher wages (ibid.).

When it comes to payment, Cynthia Hernandez and Carol Stepick (2012) published a report on wage theft in Florida, highlighting the important role that FWAF has played in seeking wages owed not only for farmworkers, but also for laborers in other industries as well. Between March 2009 and December 2010, 105 wage theft cases were reported to the FWAF, amounting to over $225,000 owed, or more than $2000 per employee. Only nine of these cases were specifically pertaining to agriculture, and the others varied across sectors like construction, other food services, transportation, retail, and healthcare. As the authors note:

> Some individuals may have felt more secure contacting an office known to serve immigrant workers than calling a government agency. Whatever the reason, the responses from workers in industries as diverse as Construction and Educational Services show how wage theft is not confined to the lowest income groups of workers, but is a business model that victimizes workers across industries and wage levels.
>
> (ibid., p. 23)

Word of the organization's reputation for assisting employees in reconciling cases has spread through different media outlets, which has not only increased the number of cases reported, but hopefully also increases general awareness of wage theft and workers' rights regarding wages.

Justice claims (recognition, capabilities, and participation)

As I move through this chapter and the case study in the next, I am building a bridge to an as-yet unexplored (in this book) facet of environmental justice as it pertains to farm labor: justice claims, or an exploration of how things should be and how farmworkers' unique strengths and capabilities are recognized and valued in authentic participation in change that impacts them. An article entitled "Florida Farmworkers Identify Issues and Priorities for Next Five Years," in *Equal Voice Magazine* in 2013 highlights the interconnected nature of the FWAF's work by discussing its statewide General Assembly with over 300 attendees and delegates to agree upon the top three issues facing farmworkers and their communities and that will direct the organization for the next five years until the next General Assembly: (1) the need for immigration reform; (2) health and safety in the workplace; and (3) economic issues. This article sums up this work stating, "The association builds power among farmworkers and rural low-income communities to respond to social, political, workplace, economic, health, and environmental justice issues that impact workers' lives" (*Equal Voice Magazine*, 2013). The Farmworker Association of Florida is an example of an organization focused on justice that places farmworkers and their voices, skills, and perspectives at the center of all they do, and in what follows I showcase some illustrations of how they do this not only effectively, but in service with farmworkers.

How should things be?

Advocacy on behalf of farmworker health and safety

The FWAF performs its advocacy at both the grassroots level as well as working on higher-level systemic change on policy and legislation (and their enactment and enforcement) impacting farmworkers. As Arcury and Marin point out, organizations like FWAF "are active in supporting new state and national legislation that promotes health, safety, and justice for farmworkers and their families" and also "work to amend existing agricultural exceptionalist laws that affect farmworker health, safety, and justice" (2009, p. 32). In addition to doing this work itself as an organization, FWAF has a long history of forming strategic coalitions with other organizations to strengthen their position on various aspects of systemic-level change. According to Palermo and Ehlers, coalitions like the ones that FWAF facilitates and participates in are "both temporary and longer-term alliances of factions for the specific purpose of enhancing health and safety of agricultural workers and their families" (2002, p. 162). As discussed previously in this chapter and reviewed by Palermo and Ehlers is the FWAF's role as a co-founder and member of the Farmworker Health and Safety Institute, which "developed and implemented the train-the-trainer program including leadership modules for migrant, seasonal workers; created links between communities and local healthcare providers" (ibid., p. 166). While this was perhaps one of the FWAF's earlier coalition-based collaborations, it set the stage for the work that the organization could do, both on the ground and from a broader perspective in its relationships with other like-minded organizations (more of which will be detailed in Chapter 6).

Currently, the FWAF is playing an integral and central role in advocating for a strong implementation and enforcement plan for the Worker Protection Standard (WPS) as overseen by the Environmental Protection Agency (EPA). Jeannie Economos of FWAF worked with Kerry Brennan of the Migrant Clinicians Network and Marco Salerno of the Farmworker Support Committee (CATA) to collect and document farmworkers' accounts of pesticide exposure on the job to submit during the public comment period of the new WPS. They worked during the public comment period to ensure that:

> Farmworkers from various backgrounds submitted public comments, met with relevant EPA policymakers such as the staff from the Office of Pesticide Programs and Administrator Gina McCarthy, and organized fellow members of their communities to become active. Farmworker advocates also worked with a coalition of supporters including clinicians, lawyers, legislators, students, faith leaders, pesticide reform activists, and, most importantly, agricultural community leaders to submit comments regarding the proposed revisions.
>
> (Brennan, et al., 2015, p. 364)

In making space for farmworkers to be involved in this policy change in this way, they gave them the "empowering opportunity to make their voices heard" while recognizing this as a first step and that what is next is "ensuring that the EPA and state governments implement and enforce a meaningful and strengthened WPS to safeguard farmworkers' livelihoods" (ibid., p. 373).

One major area of consideration with farmworker health and safety is how to change behavior on the ground—both that of the farm owners who have control over how pesticides are applied and used as well as providing access to hand washing and clean water areas, but also in the behavior of the farmworkers themselves in protecting themselves from harmful exposure and appropriately addressing exposure when it does happen. Toward that end, Antonio Tovar-Aguilar, Juan Concha, and Jeannie Economos (Tovar-Aguilar et al., 2014) presented a multi-method assessment to better understand the challenges faced by workers as they intersect with access to information, training, and safety precautions pertaining to pesticide exposure. In their work, they determined that corporate farm owners and managers did not reach the expectations set out by the Worker Protection Standard to train and provide ongoing learning opportunities for their workers in relation to pesticide use due to a lack of adoption of corporate social responsibility into the ethos and practices of many of these operations.

FWAF not only attempts to influence and change business practices, but has also worked on community-based social marketing to educate farmworkers in an attempt to reduce the harmful effects of pesticide exposure. In partnership with researchers from the University of Florida and a social marketing group at the University of South Florida, the FWAF formed the Together for Agricultural Safety project to "assist agricultural worker communities in creating effective solutions to the problem of pesticide exposure" (Flocks et al., 2001, p. 462). According to the authors of this study:

> The bedrock of the social marketing approach is extensive formative research, which involves data collection activities conducted before a program begins in order to understand how consumers perceive a product, its price, and other factors that influence consumer behavior … Research is used in social marketing to segment the target population into homogeneous subgroups based on characteristics that influence their responsiveness to marketing interventions, such as the product benefits they find most attractive.
>
> (ibid., p. 462)

All of this is done to get the target audience to change their behavior. In exploring how to apply this approach to more effectively influence farmworker behavior on health and safety, the authors emphasize the importance of each stakeholder participant group (farmworkers, academic partners, community partners) experiencing what is entailed in the others' work, learning skills and better understanding not only how to influence prevention, but why it is necessary in the first place.

In addition to social marketing, community-based prevention marketing (CBPM) is a relatively new program-planning framework that allows organizations and communities, such as the Farmworker Association to do what Paul Monaghan et al. discuss, citing Carol Bryant and colleagues (2000), using a "combination of a data-based strategic planning model with the use of diverse community members' wisdom, skills, experience, and understanding of local culture and network ties" (Monaghan et al., 2008, p. 72). Monaghan et al. suggest that this method, employed by the Farmworker Association of Florida (one of the authors of this study was Tirso Moreno, co-founder of FWAF), allows community members to build a strategic plan using five key concepts: product, competition, price, place, and promotion. Considerations of what practices will reduce or minimize harmful behavior, as well as incentives for doing so, are researched and provided through brand development and marketing to the populations most in need of the information or product. In discussing how to use community-based preventive marketing (CBPM) to improve farmworker safety, they suggest an eight-step process that includes (Monaghan, 2008, p. 73): (1) mobilize the community; (2) develop a community profile; (3) prioritize and select target behavior; (4) build community capacity; (5) conduct formative research; (6) develop a marketing plan; (7) develop program materials; and (8) implement, monitor, and evaluate the program.

In another study that the FWAF partnered on, the principles of CBPM were applied to the adoption of safety eyewear among citrus harvesters (Monaghan et al., 2011). Eye injuries are among the most common injuries for citrus workers. Debris from the trees may fall into the eyes of workers as they are reaching up to pick fruit, or eyes can be injured by pokes from branches and thorns while harvesting. After collecting baseline information on workers' perceptions of potential risks while working and on the benefits of protective eyewear, researchers followed up by observing workers in the field, conducting interviews, and preparing and implementing some targeted interventions in the form of training, educational sessions, and interceptions to encourage the use of safety eyewear. They also involved the workers in selecting the best eyewear for the job, that which was breathable in the hot Florida weather and did not distort their vision during use (those that allowed workers to maintain productivity). While working in partnership with farmworkers, they found, perhaps not unsurprisingly, that those who had adopted the eyewear and who were reaping the benefits were the most effective in influencing their peers to use this safety measure. The FWAF's interest in and commitment to continuing to find better grassroots, proactive, and preventative measures for farmworker health and safety illustrate its commitment not only to positive change for justice, but also to involve the workers themselves in coming up with those solutions.

Community activism through quilting

In addition to this important work regarding the health and safety of farmworkers, the Farmworker Association of Florida continues to seek a sense of justice for farmworkers who have been harmed over time. Creating a space for them to be recognized and acknowledged for the work that they have contributed over many years, the FWAF advocates for their lives and work to be dignified, despite their endless struggle with health issues and illness. For this to happen, the intentional creation of a space for healing provides the foundation for public participation and mechanisms for change. Without the crucial step of airing grievances and establishing the emotional realities of injustice, the path towards creating and representing that injustice is often traumatizing and can lead to more harm to those through reliving experiences. Healing as a foundational step in this process therefore better provides emotional stability, clarity, and the truth necessary for such a project. What follows in this section derives from a pilot research study conducted as part of my doctoral coursework, and is a portraiture project about Linda Lee, a former farmworker in the Apopka, Florida, community who had a vision to facilitate the healing process in order to allow farmworkers to tell their stories of injustice through quilting (Berkey, 2010).[2]

To those in the Apopka community, Linda Lee is known as a quiet, strong woman, unwavering in her love for children and steadfast in her own fight against lupus and other chronic illnesses. My students lean forward on the edges of their seats for once, not because they are antsy to go but because, in her soft, accented voice, Linda holds up her hand-made works of art, one by one, and talks about the members of her family that she has loved and lost. Laced throughout these stories and the beautiful folk art pieces is her will to survive, and to pay homage to the people who meant the world to her. Today, it is a group of college students looking to go into business for themselves some day. The other day, she and several other former farmworkers and members of the Farmworker Association of Florida were in Indiantown, another farmworker community three hours from Apopka, with fabric strewn about in an attempt to inspire others to begin to heal by telling their own stories. In tireless bouts of education and inspiration, she encourages people to find their own voices, empowering them to take control of their own destinies and challenging them to think about a world where it is okay to question and criticize the status quo unselfishly, for the benefit of others. While she talks to people from different backgrounds and even generations, it is clear to see that her ability to act as a human bridge allows for an intergenerational exploration of morality, ethics, and compassion (Page, 1999).

Not only this, but Linda and some of her closest friends, including Geraldean whom I discussed earlier this chapter, struggle each day to get out of bed. It is no longer just speculation that the pesticides they have all been exposed to since childhood, as they worked long hours in the fields and came home to places in a farming community exposed to the elements, have harmful health effects on people as well as the environment. As she describes it:

> Now we got youth, we got young ladies having babies that are breaking out in rashes they call eczema, it looks like a bubble on the top of they skin. And you know, it makes me cry to see a little child, 2 and 3 years old

> in so much pain, so. With the stuff that we going through, me and Geraldean—she is on dialysis, I been on peritoneal dialysis and I did my own stuff, thank God, in 2005 I received a kidney, but Geraldean she is on the list and she trying to give up but I'm trying to keep her encouraged because of you know, we have to keep trying to fight to stay alive to educate our young people you know that if you go in these nurseries, on the farm, or wherever you go, and they be using chemicals, that you need to be using protection.

Looking around the Farmworker Association of Florida, one might think that little warmth exists within its white walls, with basic furniture and a box-like structure. But the warmth of the place is apparent and comes from the gathering of voices like Linda's, sharing experiences of joy and pain with one another, and educating future generations about what the consequences of their actions truly are and about the reality and injustices of agricultural work. It is within this space that social capital exists, the shared knowledge, understandings, values, and other elements necessary to cultivate the capacity of all community members to reach the desired results (Dhesi, 2000).

To hear Linda and other farmworkers talk about it, there is sadness in the feeling of being left behind, but she has a special knack for appreciating what she's been given, she is full of gratitude, and shows her support for the community by doing something some people would say is not your typical activist approach: quilting.

When one first sees the quilts that are the result of Linda's and the community's leadership as part of and within the Association and with dedicated volunteer and artist, Sarah Downs, it is easy to be drawn into the story of what is represented there and to forget about the work that went into it. Community elders are often agents of change, leading and inspiring programs worldwide (Ross, 2003) and the Apopka area community is clearly no different. Community members, students, and others are riveted by the beauty of the hand-sewn pieces, and are full of questions about their meaning. While some quilt pieces depict deceased farmworkers doing their daily work in the fields, many of them memorialize them in a joyful aspect of their lives. Linda Lee can tell you about the inspiration for all of her pieces, such as:

> Because, like I say, with the quilt, I thought that making that piece there, that the lady that had the little pocket apron, I made that in memories of my mom. Then I thought about my grandmama, cause when she wasn't workin' on the farm, she liked to cook. She was always trying to fix something to make sure we had something, sometimes she would fix a tin can, oh, what was it, a cookie, like a snickerdoodle cookie, teacakes, that's what it was, and she would have a tin can of teacakes, she would start to cookin' em on a Friday evening and get through on a Saturday evening. So, we would come home from school, we could have teacakes and juice or whatever. She always kept something sweet, she always was doing something for us, so I did a piece for memory of her.

One finds inspiration, leadership, and vision at every turn at the Farmworker Association of Florida. Several years ago, when the AIDS quilt was touring the country, someone from the Apopka community mentioned the possibility of a memorial quilt for farmworkers who had passed away unexpectedly, many times due to pesticide-related illness. While at first the idea didn't gain much steam due to lack of time and resources to devote to it, it re-emerged some time later when a New College student in his senior year named Sam Chillaron doing an internship project with FWAF spent a summer working with the Association. He did research on quilt making, began to assemble the necessary materials and tools, and jump-started the project by making the first quilt square. Little did he know that he would pave the way for the potential for healing and hope in this community, and for Linda. An avalanche of support poured out, and Sarah Downs was approached to work as a partner to Linda and others for the project. Much of this was catalyzed by Jeannie Economos, the Pesticide Health and Safety Coordinator at the Farmworker Association, who was working with Sam, and was interested in helping the community realize this re-emerged vision. In addition, Jim Berry contributed by doing oral histories with the farmworkers while they were doing the quilting.

However it began, the quilt project took on a life of its own as it unfolded. Its meanings multifaceted and the space it created for people to gather, reflect, and remember valuable beyond belief. When reflecting upon the personal impact of the quilt project on her life, Linda Lee stated:

> The personal impact that it had for me is because my entire family worked on the farm. We was what you call a close-knit family, my grandfather and grandmother did not want us to go anywhere but, you know, just stay together. And I guess after he died in 63, the family went to Carolinas, not leaving this area, but we branched out in the Apopka area where we wasn't close anymore. But on the Sundays we would always meet. So, when we would meet up we would have a lot of time to spend together, we would play softball together, we ate together, we talked amongst each other, what farm we was working on and what kinda work was going on out there and different things. Even how they treated you and different stuff. After, um, my, like I say, after my sister Margie died, it really hurted me. It really hurted me. I just made up my mind then it's now or never. We have to do something. So, when Sarah Downs came in and she said she was willing to work with me on the quilt, I said, I don't know how to do it, I'm telling you, Becca, I went to praying. And I asked God, show me a way to portray my people, you know. And, to bring this quilt out and make it look decent.

Linda Lee is among a contingent of women who seek their faith and reassurance from God through prayer, the strength to navigate the complex and muddied waters of life, and from which one has promise in her stance (Rose, 1987). From the beginning, there were different ideas about what the quilt should look like, and even about what its purpose should be. Sarah Downs talks empathetically about Linda's role in realizing the vision of the project:

> What keeps going through my mind, initially everyone has their own vision of what the quilt should look like, but this is really Linda's baby, yes, not Jeannie's or mine. You know Jeannie's background is the revolutionary anarchist, she wanted more of that, like, stick it to the man. And, me, I kind of have this art background and I'm thinking it needs to be aesthetically pleasing, and then Linda, you know, I gotta hand it to her, Linda waits until the mood strikes her and then she'll kinda work with imagery, and she said that this is important to her because this is honoring her people. I'm sorry, I'm getting … and she really knew we were pushing her a different direction and she made sure … I'm sorry (crying) … she made sure that she really stayed true to what she believed is important. She got very adamant about that, and I really respect her for that.

Linda's leadership, along with her vision and ideal for the project, as someone who had lived in this community all of her life and worked the land alongside both her own family members and others, shaped not just the final product, but the journey of the making of the memorial quilts. In close-knit communities like this one, social capital is not only necessary, but vital in enabling community members to see themselves reflected in a project (Dhesi, 2000). Linda's leadership shines through in her interactions with people from all generations, as she seeks the connection with the people she is with, rather than searching out the differences. I saw this first-hand in her ability to relate to my students, who are two generations removed, and Sarah explains her ability to connect:

> Well, we went to Indiantown several times because there is a large community there of former muck farmworkers and Linda was so helpful with, I think, making people feel at ease and comfortable, because again, she has a shared experience. In fact she's even a relative of one of them, and they'd be talking about "my cousin married your cousin," and there was a commonality, and she made a point, we would talk about it afterwards, on the way home. There was a little group of us that would drive together up to Indiantown, I mean, because she picked up on it pretty quickly that her role, she figured out, was to walk around and talk to the folks about what they were working on.

It was in this and many other ways that the space Linda created through the making of the quilts and the telling of these stories somehow paved the way between systemic mistreatment of a people and the road to healing, bringing back the memories of those who came before and sharing them with future generations (Middleton, 2010).

Many community members were uncertain about getting involved in the project, and Sarah Downs was very aware of how responsive people became once Linda talked with them about their stories and making their personal quilt pieces. She talks further about their time working with farmworkers directly, stating:

> I think initially I was bringing books with quilts and quilt making because I'm not a quilt maker, and I was showing imagery and I would bring a book of paintings by African-American artists, and I would tell them they could transfer those ideas, and they were very polite, and I could sense I wasn't quite making that connection. And then Linda came in and said, "I was watching the kids last night and I was thinking about my mother and I did this. Look, what do you think?" And they go, "Wow, that's great. I love that. I get that."

The next steps in the journey are unclear, but one thing is for certain: the quilt is having a number of impacts, all positive, some intended and some unintended. It has already traveled around to schools, conferences, and museums for installations, and serves to educate people about the hardships faced by a sometimes forgotten segment of our society, the farmworkers who pick our food and cultivate our land. More than that, it humanizes farmworkers as the people they were in their best moments, enjoying the things that all people love. Perhaps unintentionally, the quilt project has rejuvenated a community and allowed them to honor the past, present, and future and to begin to envision what healing means. One thing everyone agrees upon is that this would not have happened without Linda Lee.

The Farmworker Association of Florida: an exemplar of justice for farmworkers in large-scale conventional agriculture

All of the efforts highlighted in this chapter contribute to the type of community-driven change that Kline and Newcomb (2013) discuss, noting that multiple efforts on behalf of the FWAF organization and its members to raise awareness of their situation have a ripple effect on students, activists, politicians, and consumers. In essence, it sheds light on a little-noticed aspect of our food system as we eat at restaurants and shop at grocery stores: the farmworkers. It gives people the sense of the injustices inherent in the structure of our system, leaving these farmworkers with little hope unless there is structural, systemic change and increased public and political will to insist on those changes. They note:

> The Apopka farmworkers worked long, grueling hours in search of a livable wage, and often brought their children to work alongside them in order to maximize their household earnings. Students are able to see how, in this scenario, the poor are an example of the hardest working people, but they lack benefits that the middle class usually associates with employment, such as healthcare. Furthermore, this allows students to begin to understand how healthcare becomes out of reach as it becomes suspect

> to neoliberal agendas. As healthcare becomes increasingly commoditized, it becomes more out of reach for many marginal populations, including the former farmworkers of Apopka.
>
> (ibid., p. 174)

It has been my unique privilege to be introduced to the hidden world of farm work in this country through working with the FWAF, and as an organization and the population it serves, it highlights the social, environmental, and structural dimensions of the justice issues impacting farmworkers at this scale of agriculture, as well as providing an example of justice claims and work that can be done in community with farmworkers to create a world where things are as they should be: with farm work being safe, dignified, and recognized for its vital role in feeding the people of the United States.

Notes

1 The book *Fed Up: The High Cost of Cheap Food* (Slongwhite, 2014) eloquently tells the stories of many of these farmworkers, and is recommended reading for anyone who would like more detail or insight into their lives. *Pesticide Lake*, a short film by Sanjay Rawal (2015) is also useful for getting a glimpse into what has happened around Lake Apopka.

2 To learn more and see the beautiful visual images of the quilt project, see Chapter 13 of *Fed Up: The High Costs of Cheap Food* (Slongwhite, 2014) and the Lake Apopka Memorial Quilt blog at http://apopkaquiltproject.blogspot.com/.

References

Arcury, T. A., & Marín, A. J. (2009). Latino/Hispanic Farmworkers and Farm Work in the Eastern United States: The Context for Health, Safety, and Justice. In S. A. Quandt, & T. A. Arcury (Eds.), *Latino farmworkers in the Eastern United States* (pp. 15–36). New York: Springer. Available at: www.springerlink.com/content/n796887132220894/abstract/

Arcury, T. A., Quandt, S. A., & McCauley, L. (2000). Farmworkers and Pesticides: Community-Based Research. *Environmental Health Perspectives*, 108(8), 787–792. Available at: http://doi.org/10.1289/ehp.00108787

Balogh, C. (2011, June 1). Apopka Farmworkers Say Pesticide Exposure Caused Illnesses. *Orlando Weekly*. Available at: www.orlandoweekly.com/orlando/apopka-farmworkers-say-pesticide-exposure-caused-illnesses/Content?oid=2248681

Berkey, R. (2010). Leadership of an Elder, the Healing of a Community: Linda Lee and the Apopka Farmworker Memorial Quilt Project. Unpublished, Antioch University New England, Research Methods II.

Berkey, B. (2012). Environmental Justice from the Ground up. In D. R. Gallagher (Ed.), *Environmental leadership: a reference handbook* (pp. 328–335). Thousand Oaks, CA: Sage.

Bermudez, L. (2012). Farmworkers in Florida: Silence Is Their Rational Choice. Tallahassee, FL: Florida Association of Community Health Centers. Available at: www.fachc.org/pdf/Farmworkers%20in%20Florida_Silence%20is%20their%20rational%20choice.pdf

Brennan, K., Economos, J., & Salerno, M. M. (2015). Farmworkers Make Their Voices Heard in the Call for Stronger Protections from Pesticides. *NEW SOLUTIONS: A*

Journal of Environmental and Occupational Health Policy, 25(3), 362–376. Available at: http://doi.org/10.1177/1048291115604428

Bryant, C. A., Forthofer, M. S., Brown, K. R. M., Landis, D. C., & McDermott, R. J. (2000). Community-Based Prevention Marketing: The Next Steps in Disseminating Behavior Change. *American Journal of Health Behavior*, 24(1), 61–68. http://doi.org/10.5993/AJHB.24.1.9

Central Florida Foundation. (2014). Farmworker Association of Florida, Inc. (FWAF). Available at: http://cfcf.guidestar.org/FullPDF.aspx?orgId=2569&preview=False

Comas, M. E. (2015). Sick Apopka Farmworkers Ask for Research on Medical Ills. *Orlando Sentinel*. September 19. Available at: www.orlandosentinel.com/health/os-apopka-farmworkers-lupus-20150918-story.html

Dhesi, A. S. (2000). Social Capital and Community Development. *Community Development Journal*, 35(3), 199–214. Available at: http://doi.org/10.1093/cdj/35.3.199

Earthjustice, & Rawal, S. (2015). Pesticide Lake: The Poisoning of the Workers of Apopka, Florida. Available at: www.youtube.com/watch?v=vA2WiQwJ3xI

Equal Voice News (2013). Florida Farmworkers Identify Issues and Priorities for Next Five Years. *Equal Voice News*. March 19. Available at: www.equalvoiceforfamilies.org/florida-farmworkers-identify-issues-and-priorities-for-next-five-years/

Estabrook, B. (2012). Of Money and Muck. *Gastronomica*, 12(1), 30–33. Available at: http://doi.org/http://dx.doi.org.ezproxy.neu.edu/10.1525/GFC.2012.12.1.30

Farmworker Association of Florida. (2015). *Campesino a Campesino Encuentro (Final Report 2015)*. Apopka, FL: Farmworker Association of Florida.

Farmworker Health and Justice Team. (n.d.). Coming Clean, Inc. Available at: www.comingcleaninc.org/programs/farmworker-justice (accessed April 25, 2016).

Flocks, J., & Burns, A. (2006). Stakeholder Analysis of Florida Farmworker Housing. *Journal of Agromedicine*, 11(1), 59–67. Available at: http://doi.org/10.1300/J096v11n01_07

Flocks, J., Clarke, L., Albrecht, S., *et al.* (2001). Implementing a Community-Based Social Marketing Project to Improve Agricultural Worker Health. *Environmental Health Perspectives*, 109 (Suppl. 3), 461–468.

Florida Senate. (2011). Status of Health-Related Consequences to Muck Farm Workers in the Lake Apopka Region (Issue Brief No. 2012–2201). Available at: http://wayback.archive-it.org/4495/20140329230351/http://www.flsenate.gov/PublishedContent/Session/2012/InterimReports/AG2012012-201ag.pdf

Guillette, L. J., Gross, T. S., Gross, D. A., et al. (1995). Gonadal Steroidogenesis in Vitro from Juvenile Alligators Obtained from Contaminated or Control Lakes. *Environmental Health Perspectives*, 103, 31–36. Available at: http://doi.org/10.2307/3432409

Habin, R., & Matthew, G. (2006). *Lake Apopka farmworkers environmental health project: report on community health survey*. Apopka, FL: The Farmworker Association of Florida. Available at: http://static-lobbytools.s3.amazonaws.com/press/20110713_lake_apopka.pdf

Hernandez, C. S., & Stepick, C. (2012). Wage Theft: An Economic Drain on Florida. Florida International University: Research Institute on Social and Economic Policy: Center for Labor Research and Studies. Available at: www.floridafarmworkers.org/images/my_images/pdf/wage_theft.pdf

Kamel, F., Moreno, T., Rowland, A. S., *et al.* (2001). Recruiting a Community Sample in Collaboration with Farmworkers. *Environmental Health Perspectives Supplements*, 109, 457.

Kline, N. (2010). Disparate Power and Disparate Resources: Collaboration Between Faith-Based and Activist Organizations for Central Florida Farmworkers. *NAPA Bulletin*, 33(1), 126–142. Available at: http://doi.org/10.1111/j.1556-4797.2010.01044.x

Kline, N., & Newcomb, R. (2013). The Forgotten Farmworkers of Apopka, Florida: Prospects for Collaborative Research and Activism to Assist African American Former Farmworkers. *Anthropology and Humanism*, 38(2), 160.

La Asociación Campesina de Florida/Farmworker Association of Florida. (n.d.a). Immigrants. Available at: www.floridafarmworkers.org/2011-08-08-16-02-46/immigrants-rights (accessed April 25, 2016).

La Asociación Campesina de Florida/Farmworker Association of Florida. (n.d.b). Asosiyasyon Travayé laté. Available at: www.floridafarmworkers.org/ (accessed April 12, 2016).

La Asociación Campesina de Florida/Farmworker Association of Florida. (n.d.c). Our History. Available at: www.floridafarmworkers.org/about-us/our-history (accessed April 12, 2016).

La Asociación Campesina de Florida/Farmworker Association of Florida. (n.d.d). Pesticide Actions. Available at: www.floridafarmworkers.org/2011-08-08-16-02-46/pesticide-actions (accessed April 25, 2016).

La Asociación Campesina de Florida/Farmworker Association of Florida. (n.d.e). Workers Justice. Available at: www.floridafarmworkers.org/2011-08-08-16-02-46/workers-justice (accessed April 24, 2016).

Lake Apopka Farmworker Memorial Quilt Project. (n.d.). Available at: http://apopkaquiltproject.blogspot.com/m (accessed April 10, 2016).

Mayer, B., Flocks, J., & Monaghan, P. (2010). The Role of Employers and Supervisors In Promoting Pesticide Safety Behavior Among Florida Farmworkers. *American Journal of Industrial Medicine*, 53(8), 814–824. Available at: http://doi.org/10.1002/ajim.20826

Middleton, E. (2010). A Political Ecology of Healing. *Journal of Political Ecology*, 17, 1–28.

Monaghan, P. F., Bryant, C. A., Baldwin, J. A., *et al.* (2008). Using Community-Based Prevention Marketing to Improve Farm Worker Safety. *Social Marketing Quarterly*, 14(4), 71–87. Available at: http://doi.org/10.1080/15245000802477607

Monaghan, P. F., Bryant, C. A., McDermott, R. J.*et al.* (2011). Adoption of Safety Eyewear Among Citrus Harvesters in Rural Florida. *Journal of Immigrant and Minority Health*, 14(3), 460–466. Available at: http://doi.org/10.1007/s10903-011-9484-3

Page, E. (1999). Intergenerational Justice and Climate Change. *Political Studies*, 47(1), 53.

Palermo, T., & Ehlers, J. (2002). Coalitions: Partnerships to Promote Agricultural Health and Safety. *Journal of Agricultural Safety and Health*, 8(2), 161–174.

Permanent People's Tribunal. (2011). The People's Tribunal Session on Agrochemical Transnational Corporations. Pesticide Action Network International. Retrieved from www.multiwatch.ch/cm_data/Anklaggeschrift_TPPBangalore_2011.pdf

Permanent People's Tribunal. (n.d.). Session on Agrochemical Transnational Corporations Case Summary.| Available at: www.agricorporateaccountability.net/en/page/general/20 (accessed July 25, 2016).

Protect Farmworkers. (n.d.). About Us. Available at: http://protectfarmworkers.org/suborganizations/

Rose, S. D. (1987). Women Warriors: The Negotiation of Gender in a Charismatic Community. *Sociology of Religion*, 48(3), 245–258. Available at: http://doi.org/10.2307/3711521

Ross, F. (2003). Call to Action to Elders Worldwide: The Need to Highlight Their Abilities, Wisdom, and Compassion as Citizen Diplomats, and Leaders for Social Change. *Peace and Conflict Studies*, 10(1), 116–123.

Slongwhite, D. F. (2014). *Fed up: the high costs of cheap food*. Gainesville, FL: University Press of Florida.

Tovar-Aguilar, J. A., Concha, J., & Economos, J. (2014). Farmworkers' Health and Safety Training by Grassroots Organizations: The Importance of Greater Corporate Social Responsibility to Protecting Workers' Health. *Journal of Agromedicine*, 19(2), 244. Available at: http://doi.org/10.1080/1059924X.2014.892856

5 Case study in organic agriculture

The Northeast Organic Farming Association

Introduction to the Northeast Organic Farming Association

As Chapter 4 has detailed, I had my first foray into farmworkers' issues through my collaboration with the Farmworker Association of Florida (FWAF). It is important to note that FWAF is an organization focused on farmworkers' rights and issues, and while there are other organizations that have this focus (as showcased in Chapter 6), it is also common for farmworker issues to be among the many agenda items for organizations focused on agriculture more broadly. The case study for this chapter on the Northeast Organic Farming Association differs from that in Chapter 4 on the FWAF not just in its focus on the other end of the continuum on small-scale, organic agriculture, but also because farmworker and farmer justice are one of many issues on which this organization focuses. Therefore, while there are process, evidence, and justice claims illustrated by this case study, they in and of themselves are not the explicit reason for the organization's existence.

This chapter is also a unique case study given my interactions with NOFA during my academic research.[1] While in my doctoral program, and spurred on by the work I did with the Farmworker Association of Florida (FWAF, see Chapter 4), I knew I cared about three things: (1) farmworker issues; (2) environmental justice; and (3) how people can make changes regarding issues that matter to them. My doctoral program included a 300-hour service project, and when I came to the point in the program where I needed to pick a site for this, I wanted to work with an organization that touched on farmworker issues and justice, and that operated at a regional scale minimally. Through Jeannie Economos at the FWAF, I was connected with a farmworker activist named Richard Mandelbaum, who at the time was serving as the chair of the board for the Domestic Fair Trade Association (DFTA), a national initiative with members from across five sectors of the economy, including growers, buyers, sellers, NGOs, and farmworker organizations, that is committed to translating the best practices of the international fair trade movement to the domestic market. I signed on to do my service project with them, beginning the work of translating their principles into a rubric by which they and others could evaluate fair trade labels used in the domestic market (more on this in Chapter 6). These principles fell predominantly into three areas: health, justice, and sustainability. My learning curve was steep because I did not have a background in fair trade and its principles, but I learned a great deal from graciously patient mentors. I had the opportunity to attend their annual meeting in December 2011, where I got to meet in person many of the change agents who are a part of their organization, and where I initially connected with the Northeast Organic Farming Association, which is part of the DFTA. Through my connections in this network, I was introduced to Liz Henderson from the Interstate Council of the Northeast Organic Farming Association (NOFA).

A farmer for over 30 years, Liz was instrumental in the founding and success of NOFA. I went into our initial conversations curious about her thoughts on environmental justice issues as they impacted workers on organic farms, since most of my scholarship and experience to that point had been about farmworkers in conventional agriculture, where the predominant researched issue is pesticide-related illness. I knew that a central tenet of organic agriculture was that pesticides were not part of the production process, so I naively wondered, what could the issues be? Pesticides aside, it became clear in my first conversation with Liz that issues such as substandard housing, lack of benefits and health insurance, low pay, and health issues due to the physical nature of the work were just as prevalent in organic farming, particularly on the small- and family-scale organic farms, with which she was most familiar due to her work with NOFA. She made it clear that these issues impact not only the workers on these farms, but many times the farmers themselves. Small-scale organic agriculture, she educated me, occupies a difficult space in the market, trying to compete with large-scale farms that can distribute their products more widely, and lacking the governmental subsidies like those found in conventional agriculture.

In this chapter, I hope to offer some insights into the research I did with NOFA as an empirical case study of environmental justice issues as they play out in organic agriculture specifically. NOFA is a coalition of seven state chapters whose aim is "to advocate for and educate on organic and sustainable agriculture, family-scale farming and homesteading in rural, suburban and urban areas, agricultural justice and other related policy issues" (NOFA Interstate Council, n.d.). In conversation with the NOFA Interstate Council, which serves as the board of the NOFA chapters' coalition, I designed a mixed-methods study (Berkey, 2014) to both answer the research question and inform NOFA's program and policy activities. In this chapter, I will summarize relevant portions of that study. I will use the term "Northeast" throughout this chapter to refer to the northeastern region of the United States, consisting of the seven states in which NOFA operates: Connecticut, Massachusetts, New Hampshire, New Jersey, New York, Rhode Island, and Vermont.

The overarching question guiding this research was:

> How do various justice-related issues (including competition in the market, pay, housing, and health) impact farmers and farmworkers on

organic farms in the northeast, and how does involvement in a grassroots organization like the Northeast Organic Farming Association help them address these issues, if at all?

It addressed the following sub-questions:

- What are farmers' (who are participants in NOFA) perceptions of the opportunities, challenges, and pressures related to justice for farmworkers?
- How does participation in broader justice-focused or justice-related organizations affect farmers' attitudes about and actions regarding justice for farmworkers?
- What is the lived experience of farmers and farmworkers who are part of the organic farming system in the Northeast? What are their expressed needs to ensure equitable and just treatment? What assets are they able to bring to work toward their more equitable and just treatment?

The research questions drew upon the language used to describe farmers and farmworkers, who traditionally have been presented as distinct categories. However, as you will see throughout this chapter, the terms are not as mutually exclusive as traditionally presented, at least within small- and family scale organic agriculture in the Northeast. I will use the term "farmer" throughout to describe the farm owner him/herself; however, these farm owners were in no way the absentee farm owners often presented within the literature, or previously discussed and presented in other chapters in this book. They were laborers with the added dimensions of management and ownership. I will use the terms "farmworker," "worker," and "laborer" to describe those working on the farms who do not have ownership responsibilities. These farmworkers also brought valuable experience and knowledge to food production and, thus, could themselves be considered farmers. Because the terms farmer and farmworker used in the study and research questions were originally framed based upon the literature, and I communicated with participants in the study using those terms, I keep this language intact throughout what follows, with the recognition that these terms are not mutually exclusive, nor do they fully capture the nuances of reality (Berkey, 2014; Berkey & Schusler, 2016).

Situating the work in the continuum of farm labor and environmental justice claims

Before framing this study in the context of environmental justice, I will revisit the definition of organic agriculture. According to the International Federation of Organic Agriculture Movements (IFOAM), "Organic agriculture is a production system that sustains the health of soils, ecosystems and people. It relies on ecological processes, biodiversity and cycles adapted to local conditions, rather than the use of inputs with adverse effects" (IFOAM, n.d.a). IFOAM describes organic agriculture as a combination of tradition, innovation, and science, which should ultimately benefit the environment and promote fair relationships and a good quality of life for all involved. Another definition states organic agriculture as:

> a defined and certified system of agricultural production that seeks to promote and enhance ecosystem health whilst minimizing adverse effects on natural resources. It is seen not just as a modification of existing conventional practices, but as a restructuring of whole farm systems.
>
> (Morison et al., 2005, p. 25)

Despite these more holistic definitions, the United States Department of Agriculture (USDA) organic certification standards privilege ecological and environmental goals but are vague on social and economic goals. To meet the USDA's organic labeling requirements, products must be produced without excluded methods (e.g., genetic engineering, ionizing radiation, or sewage sludge), produced per the National List of Allowed and Prohibited Substances, and overseen by a USDA National Organic Program-authorized certifying agent, following all USDA organic regulations (USDA, 2012). For some farmers using organic practices, certification is not worth the financial cost or time, with advantages lacking for farmers who practice subsistence farming or do not engage in the market. The organic certification may also have no benefits depending on their interactions with buyers and their priorities for growing organically to begin with (IFOAM, n.d.b). There is a diversity in rationales that guide farmers' decisions regarding organic methods.

To understand the intersection of organic farming and environmental justice, we must return to the discussion of the environmental justice movement. As discussed in Chapter 1, Bryant's definition has been lauded as one that represents traditional notions of the field: "the systematic exclusion of people of color from environmental decisions affecting their communities" (1995, p. 5). Environmental justice expanded the focus of the environmental movement, which had previously often excluded the effects of decisions on human beings. The environmental justice field began in its early days with a focus on environmental racism, or the unequal distribution of environmental benefits and pollution burdens, particularly with respect to toxic and hazardous waste impacts on low-income and communities of color. This evolved into a consideration of broader issues of environmental inequality, such as transportation, health, housing, and energy development, and an expansion of the factors being considered, such as class, gender, age, citizenship, and their interconnections (Sze & London, 2008). Small-scale organic agriculture and its farmers and laborers are a population marginalized within the larger political-economic landscape of U.S. agriculture. Although much of the research on environmental justice issues impacting farmworkers is from conventional agriculture and therefore heavily focused on pesticide use and related illness, there are also a host of issues facing workers in organic agriculture.

Although certain types of pesticides are permissible within organic labeling, and chemical drift and contamination from nearby farms can also occur, substandard housing, lack of benefits and health insurance, low pay, and health issues due to the physical nature of the work all remain primary concerns for organic farming. Small-scale organic agriculture must compete with large-scale farms that can distribute their products more widely without the governmental subsidies like those found in conventional agriculture (Henderson, pers. comm., 2013). Therefore, discovering more about farmers' and workers' values as they relate to justice issues in organic farming can contribute knowledge to the environmental justice field on a unique and relatively unexplored marginalized population. In addition, further understanding about the motivations, drivers, and orientations toward change on the part of the leadership of this regional-scale, multi-state organizations can offer a perspective on how to influence the contexts in which organic agriculture is practiced and rewarded, therefore impacting the livelihood of laborers in this context.

The frameworks of environmental justice explored in Chapter 1 and applied throughout this book leave one asking the question, how do we re-structure social relationships to create the conditions necessary for justice and social change, particularly when everyone enters the conversation with their own diverse interests at the forefront of their thinking? Expanding the scale of focus, in addition to a systemic analysis of the causes of injustice, makes coalition building necessary to understand complex issues and create strategies for social change, allowing participants in the agricultural system to live in accordance with their values. In addition to our focus on a population not traditionally identified as marginalized, my research partnership with NOFA, a coalition of organic farmers and related stakeholder groups in the Northeast, allowed me to examine farmers' values and their relation to labor and justice within a regional scale, as well as consider forms of justice beyond distributive, and how these manifest themselves in various ways.

Process claims (social dimensions)

According to Thomas Arcury, Melinda Wiggins, and Sara Quandt, three predominant challenges are associated with understanding and adequately addressing the justice issues facing farmworkers, workers who are denied the right to organize due to the Agricultural Employee Exemption from the Fair Labor Standards Act of 1938, as discussed in Chapter 3 (Arcury et al., 2009). First, limited data exists documenting the current status of health and safety for farmworkers. Second, the data that is available indicates problems for farmworkers and their families' health and safety, particularly in the areas of housing, adequate insurance coverage, and protection and training. Third, "agricultural labor policy supports the exploitation of farmworkers, increases the risk of injury and illness, and denies justice" (ibid., p. 223). However, even with these challenges, Arcury et al. describe positive trends toward farmworker justice, as seen in the effects of the work of advocacy organizations, farm labor organizations, and the expansion of community-based research on farmworkers.

Labor on organic farms intersects with the ongoing agricultural labor dialogue more generally. As the studies on labor in organic agriculture previously discussed in Chapters 2 and 3 indicate, farmers and workers, and the political economic context within which organic agriculture exists, make clear, much more remains to be discovered with respect to justice and its ability to be realized in organic agriculture. Thus, this study examines the values that motivate farmers and farmworkers to participate in organic practices in the Northeast, the justice-related barriers they encounter, and the supports that help them overcome these barriers. What is presented in this case study builds on others (Arcury et al., 2009; Shreck et al., 2006) to expand understanding of who works on organic farms and which practices promote or inhibit justice for these workers.

Who?

This mixed methods study used two data collection methods: a survey questionnaire completed by farmer-members in NOFA, and phenomenological interviews with farmers, farmworkers, and members of the Interstate Council of NOFA. What follows in this section showcases network members and the methods by which the information presented in this chapter was obtained and analyzed.

Survey: design and administration

The survey was co-developed with input from NOFA and reviewed in the fall of 2012 by its Interstate Council. The survey included 36 items organized into three sections, inquiring about issues on the market for organic products (including where farmers sell their products and the issues they encounter, if any, with their major buyers), pay for workers, housing, policies such as unemployment insurance thresholds, membership of organizations such as NOFA, and the benefits the farmers derive from that membership. In addition, the survey asked about farmers' values and practices related to farming organically, such as whether they do so because it is a family tradition or because they uphold ideals about the environment. The survey design allowed for exploration of farmers' perceptions of the opportunities, challenges, and pressures of justice specific to organic farming. Finally, the survey included four open-ended questions regarding what supports and constraints farmers found in aligning their practices with their beliefs and values, as well as what supports and challenges they faced in the market for their products. To address potential threats to reliability and validity, survey questions, aligned with the conceptual constructs being measured, used practices of good survey design (Dillman et al., 2009), incorporated feedback based on review of a pilot survey by NOFA Interstate Council members to ensure questions were interpreted as intended, and emphasized confidentiality in the survey introduction to encourage farmers to respond honestly.[2]

Survey: estimated response rate

While it is difficult to identify with precision the overall response rate, given the need to rely on a volunteer sample, it is possible to arrive at a rough calculation based on estimates given by NOFA of the number of farmer-members to whom the survey was distributed. Per information collected by NOFA's Interstate Council, there are about 5000 members of NOFA across their network, approximately one-quarter of whom are farmers. Therefore, based on these estimates, 1250 farmers is the total number of the population from which this volunteer sample was drawn, indicating a 28.6 percent response rate overall.

Although the survey sample was not intended to be representative of all organic farmers in the Northeast, it is useful to have some sense of the extent to which the number of respondents in each state compares to the population of organic farmers in that state. Because data was unavailable from each of the NOFA state chapters on exactly how many farmer members they had, I used publicly available data from the U.S. Department of Agriculture (USDA, 2014), from which I pulled the number of total organic farmers to whom the survey would apply in each of the nine states (the seven NOFA states plus Maine and Pennsylvania, at the request of the NOFA Interstate Council). It is important to note that the USDA numbers represent *certified* and *exempt* organic farms and that some NOFA members are not certified although they do use organic practices. Therefore, at best, these numbers are estimates of the participation rate and how representative the participants were of the broader population of organic farmers in the Northeast.

Table 5.1 indicates the number of survey respondents and number of certified organic farms in each state. While the survey sample was not a probability

Table 5.1 Comparison of number of NOFA farmer-members responding to the survey and number of certified and exempt organic farms in each state

State	*# Respondents*	*Total # of certified organic farms in state*	*Percentage respondents based on total number of organic farms**
CT	16	122	13.11
MA	32	179	17.88
ME	68	517	13.15
NH	70	150	46.67
NJ	7	87	8.05
NY	118	917	12.87
PA	18	679	2.65
RI	7	24	29.17
VT	21	542	3.87

Note: * Sample was not a probability sample drawn from this population but a volunteer sample of NOFA farmer-members in each state.

sample, comparing these figures suggests what proportion of organic farms in each state is captured among survey participants. This comparison suggests that in some states, including Rhode Island, Connecticut, Maine, New York, and Massachusetts, the survey respondents—although not a representative sample—reflect between 12 and 29 percent of the organic farms in that state. On the other hand, the percentage of survey respondents in Pennsylvania, Vermont, and New Jersey is so low when compared to the total number of certified organic farms that it cannot be concluded that they adequately reflect the experiences and attitudes of the organic farmers within that state.

Laborers on organic farms in the northeast

Respondents were asked

> Please tell us how many people worked on your farm and were [PAID] [NOT PAID] for each category in the 2012 calendar year. "Year Round" is anyone who is a 12-month employee of your farm and "Seasonal" applies to anyone working less than that. If no one in that category worked on your farm in 2012, please enter 0.

Table 5.2 indicates the total number and mean for each type of worker reported. As these tables demonstrate, many farmers depend largely on unpaid workers, namely in the form of seasonal volunteers and customers/CSA members. The survey did not ask how many hours per week or season each type of laborer contributed, the size of the farm, nor the products produced; therefore, comparisons between worker types using these variables are not possible. The labor of family members is predominant, followed by paid employees, volunteers, interns, neighbors, and customers and/or CSA members. Many times these worker types were not mutually exclusive, meaning that workers may fall under several categories (such as family member and volunteer). Unpaid laborers make up more than twice the number of paid laborers on these farms (Berkey, 2014; Berkey & Schusler, 2016).

Site visits and interviews

Sampling and data collection

Farmers and farmworkers were chosen for the site visits and phenomenological interviews using purposeful sampling (Creswell, 2009). Purposeful sampling, in contrast to probabilistic sampling, involves "selecting information-rich cases for study in depth" (Patton, 2002, p. 169), specifically when the researcher wants to understand those cases without wanting a generalization to the entire population under study. For this study, we sampled by case type using reputational case sampling, in which one obtains the recommendation of knowledgeable experts for the best examples (McMillan & Schumacher, 1997).[3] This

Table 5.2 Total number and mean number per farm of laborers by type, paid laborers (N = 357)

Labor type and time on farm	*Total paid labor*	*Mean paid labor*	*Total unpaid labor*	*Mean unpaid labor*
Full-Time, Year Round	400	1.33		
Full-Time, Seasonal	204	0.66		
Part-Time, Year Round	203	0.71		
Part-Time, Seasonal	373	1.12		
Family Members, Year Round	156	0.54	228	0.74
Family Members, Seasonal	134	0.48	197	0.66
Interns/Apprentices, Year Round	28	0.07	16	0.05
Interns/Apprentices, Seasonal	105	0.33	97	0.34
Neighbors, Year Round	11	0.04	16	0.06
Neighbors, Seasonal	85	0.31	148	0.52
Customers/CSA Members, Year Round	302	1.12	247	0.86
Customers/CSA Members, Seasonal	423	1.6	2394	8.23
Migrant Workers, Year Round	0	0	0	0
Migrant Workers, Seasonal	29	0.11	0	0
H2A Workers, Year Round	2	0	0	0
H2A Workers, Seasonal	4	0	0	0
Volunteers, Year Round			507	1.73
Volunteers, Seasonal			1730	6.18
TOTAL	2459		5580	

purposeful selection is designed "with the intention of maximizing the utility of the data for the research goals" (Locke et al., 2000, p. 100).[4]

In this study, there might have been several criteria to use, including those farmers who indicated that they employ migrant and/or H-2A workers, those facing particular challenges in meeting a bare minimum standard for payment, compensation, and coverage for their workers, and/or those who indicate broad or deep participation in NOFA and other organizations with an interest in justice-related issues. After analyzing the data from the survey, we decided to sample for the latter type of farm, those involved in NOFA and known for their commitment to justice and related principles, because these sites could also provide best practices for use by other farms. Recommendations were collected from the Interstate Council, each of the 13 possible participants were farms contacted via email and phone when necessary. Three farms emerged as willing participants, at which I conducted site visits and interviews in August and September 2013. I interviewed a total of four farmers and six workers. For the purposes of confidentiality, I refer to the three farms as Farms A, B, and C. At Farm A, located in Massachusetts, I interviewed two farmers and one of their workers; at Farm B in New York, I interviewed one farmer and three workers; and at Farm C in Rhode Island, I interviewed one farmer and two workers. Descriptions of the farms here are pulled largely from their publicly available information, but are not cited directly to protect the confidentiality of the participants. Three of the members of the Interstate Council agreed to be interviewed, and these interviews took place in December 2013. They represent the New York, Vermont, and New Hampshire chapters, and all three interviews took place via telephone.[5]

Farm A

The farmers at Farm A are known to hire former inmates, providing jobs for individuals who would have a hard time finding employment. In addition, they put a lot of effort into training, paying living wages, and providing emotional support. Farm A has been in existence since 1982 and has been selling to the public since 1985. In addition to 2 tillable acres for fruit and vegetable growth and 100 fruit trees, Farm A also raises 500 meat birds, 13 pigs and 2 meat calves on its surrounding pastures. It was first certified organic by the Massachusetts chapter of the Northeast Organic Farming Association in 1987. Since 2003, the certification has been handled by the Baystate Organic Certifiers.

Farm B

Farm B openly states its commitment to the dismantling of oppressive structures that degrade our food system. It is a 5-acre, Certified Naturally Grown, fair trade, family farm that produces produce, eggs, and meat for diverse communities. They draw upon the wisdom of their ancestors, upholding ethical and ecological standards in land use and business practices, and acting in solidarity with people marginalized in the food system. They focus on education and community building where people gather to share skills on farming, forestry, and natural building, innovative land use practices, and contribute to the movements for justice and sustainability.

Farm C

Farm C is an urban community health and herbal education center with extensive medicinal herb gardens. The herbs for their products are grown on a 5-acre farm outside of the city. Their urban medicinal herb gardens were founded in 2000 and their herb shop/community health center opened in 2008. Their urban space is home to their teaching gardens, their store, and the health and education center. This space also hosts their production kitchen. In 2011, they purchased the farm outside the city, where they have honeybees, fields of medicinal herbs, and a classroom space. Their education programs focus on herbal information, nutritional healing, and organic farming practices. Since 2000, they have been offering free and sliding scale classes at elementary schools, nursing homes, and non-profit organizations. They also offer educational programs for colleges, hospitals and garden clubs.

Why is this happening and how did it come to be?

Here I note the themes that emerged from my interviews with three members of the Interstate Council, which are useful in discovering the overall perspective and investment of the Northeast Organic Farming Association in increasing justice for both the farmers and workers involved in organic agriculture. Many of the themes that emerged from these interviews tied into those that came from the farmers and workers themselves from the site visit interviews presented later in this chapter, although Interstate Council members reflected in slightly different ways about the actual and possible roles that NOFA can play in creating the context for justice. Overall, they discussed the themes presented in Figure 5.1.

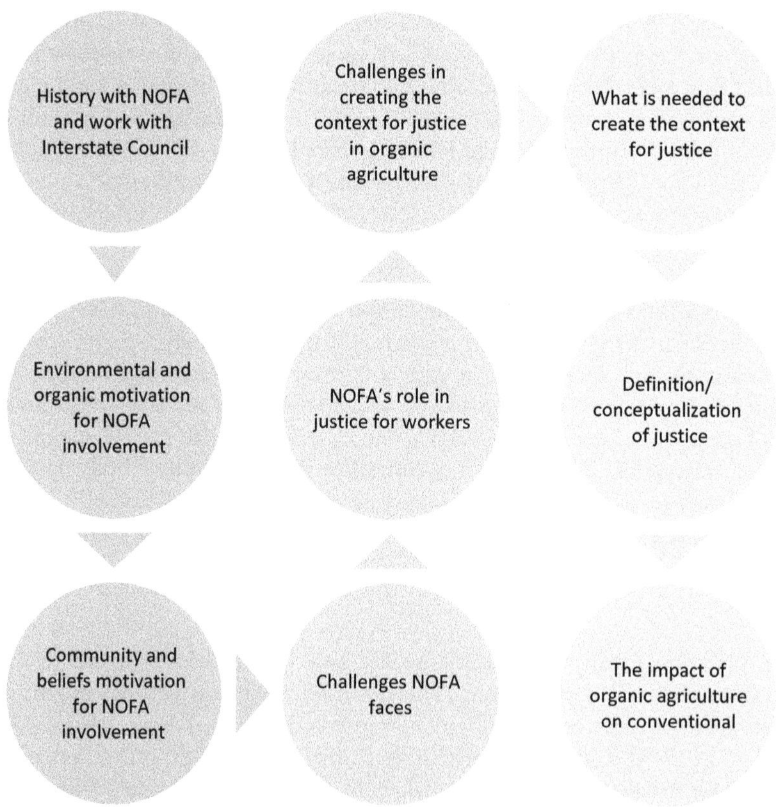

Figure 5.1 Themes from interviews with the Interstate Council members

One of the Interstate Council members interviewed, known for her activism in justice, reflected on her history with NOFA:

> Being a voice within NOFA and within the organic movement for finding ways to keep that balance, where we get the money that we need while reaching out to lower-income people and lower-income communities, has been one of the roles that I have played. And that's been really important to me.

Another member noted the work of the Interstate Council and its importance and potential for making change:

> We have the potential to really have a big voice in policy issues that sort of are national in scope, I would say. And also a tremendous opportunity to help one another with each other's problems, sharing … we're moving in that direction …

NOFA's influence of and participation in the original conversations about the organic movement, definitions, and certification requirements showcase this potential. According to one participant:

> So, in preparation for that meeting, we were going to be working on a national definition of organic, and we wanted to see if the many different organizations that were working around the country and doing certification were close enough in what we were doing that we could work together. So, several of us read through the organic standards of all of the different groups. And we discovered that we were 90 percent identical, and that the 10 percent that was different weren't differences in values, but differences in procedures.

It is in that 90 percent overlap that NOFA gathers its strength, in addition to the similarity in values among its members, according to the members of the Interstate Council interviewed. This sentiment is exemplified in the following reflection:

> The work that the organization does is consistently their own vision of the way things should be and the way things should develop and consistent with their personal values … I think that's what people get out of joining NOFA … and also if you're a farmer or consumer, you get information, I mean, we do a lot of program work; a lot of technical workshops; a lot of sharing of information. Practical skills … that's another thing that NOFA provides. Also … to help them have a voice. And to join with others and leveraging that voice in the halls of power, wherever that might be.

However, while all three participants saw great potential with NOFA in addition to their lived benefits, they also saw challenges facing the organization very clearly.

One consistent theme in this area was the need to get more young people involved in their work and in the work of organic farming. One participant states:

> Getting a younger generation involved and active in volunteering to do things. So I think the biggest challenge is … to pass that torch on; to engage youth, and get them active in the organization; to take leadership roles, volunteerism, that's the only way the movement will survive. I see some good signs of that, but that is a challenge. And … more diverse groups. So, we have as a core organic farmers and gardeners. But increasingly we need to appeal to young parents and land care providers, and just foodies. People that want to support organic food, because they like to eat organic food.

Another tension is that between the power and funding of the state chapters versus the Interstate Council. There were remarks concerning how much the state chapters truly want to work together rather than independently, which has implications for the ability of this network to support justice-based (and other) initiatives and education.

When it comes to justice for farmers and workers within their network and on organic farms in the Northeast, the participants had different views on what this could look like, in addition to what NOFA's role could or should be in helping that vision happen. One participant stated that the diversity of worker and farm types within the network alone makes it difficult to conceptualize what a universal definition of justice and its supports could be, pointing out:

> I mean, there are large farms; small farms; there are internships; apprentices; there are seasonal workers; there are long-time workers who are actually like part of the family, you know. Everything, migrant workers, and so, I don't know if there's any. I think the, the sort of conceptualizing what would be fair in all those situations and what the structure of that might look like. I don't know, I guess I can't get my head around that.

On the flip side, another participant points out that with the appropriate commitment, this same diversity mentioned as a challenge can also become a strength, stating:

> And to strengthen that collaboration, because we've got a lot of people in the Northeast. There are a lot of people living in these NOFA states, and if we really figure out how to do a better job of representing those voices of all those folks, and getting people to join, and getting people to think about agriculture, and getting people to realize what stake they have in it.

All of the participants mentioned the need for justice-related certification to make sustainable change, saying that certification criteria and accountability create the validation and legitimacy needed to really create change. One combines this with the importance of communication:

> So what NOFA has to do is support those type of programs and certifications through its communication vehicles. Through the state chapter newsletters. through the Interstate Council website. We need to communicate, and get the message out so we can have those workshops take place and certifications. And get certifications, much more so than they're happening. I know that moves slow, but that's what NOFA can do.

A participant describes NOFA's role as not just about certification, but also about the networked ability to organize farmers around a common cause and to unite their voices for positive change that would affect the whole system-including farmers, workers, the earth, animals, etc.:

> Well, in an ideal world, NOFA would be instrumental in helping the farmers organize so that they could get the prices that they need to all get living wages and pay their workers living wages. And NOFA would be part of a movement that would raise the minimum wage so all working people could afford to buy the food that we grow.

Expanding on this, another participant discusses the market pressure that small-scale organic agriculture finds itself in, stating:

> Then the prices that stores pay are not enough to cover the cost, the real cost of production. I mean, you probably understand about that, but there are all these externalities like, you know, the contamination of the soil and the water and the air that happen from conventional agriculture. And the illnesses that happen to people when they eat low-quality food. That's not part of the price that people pay in the store. So in organic production we are not ringing up those costs. And organic production requires more management, more attention, more care. And so it is more expensive than vast monoculture. It's higher quality. It's higher price.

Summing up the themes that emerged throughout my conversations with the Interstate Council participants, one states their perception of the central purpose of NOFA's work:

> Taking care of each other. Providing healthful food. I mean, all of that is, should be at the very top of our sense of what we value and what we reward. But that's not the way we look at it. I think that's a sort of a overarching kind of a view of what justice might require.

Evidence Claims (Environmental/Structural Dimensions)

This study, including the survey and interview results that follow, is useful in highlighting the structural dimensions of injustice faced by farmers and laborers on organic farms in the NOFA network, including what is happening with the misalignment of their goals and motivations and the political economic context within which these farms operate (Berkey, 2014).

Survey findings: farmer value ratings

Respondents were asked to rate a series of items based on their importance in the farmer's decision-making in and rationale for organic farming. In other words, they rated each item on its level of influence on them as a farmer. For each of the items, a range of 270–284 respondents (N) rated the item on a scale of 1–5, with 1 being of least importance and 5 being of most importance. Table 5.3 presents the mean rating of these values and provides the standard deviation for each.

Survey findings: supports for and constraints in implementing values

Six major themes emerged from open-ended responses to the question "What supports enable you to implement farm practices that align with your priorities as rated here?" (see Table 5.3 for ratings). These were: (1) NOFA and MOFGA

Table 5.3 Values including standard deviation for farmer value ratings

Value	*N*	*Mean rating*	*Standard deviation*
Producing high quality products	283	4.71	0.77
Providing highly nutritious food	284	4.67	0.84
Avoiding chemicals for environmental health	281	4.67	0.87
Avoiding chemicals for family/farm worker health	282	4.67	0.83
Improving or maintaining land stewardship, ecological sustainability, ecological principles	282	4.57	0.87
Improving or maintaining ethical trade relationships (fair, honest, transparent)	277	4.33	1.03
Supporting the buy local movement	279	4.1	1.11
Providing a fair/just workplace for farmworkers	272	3.89	1.28
Reducing input costs	278	3.86	1.09
Sharing health and safety issues with your employees	270	3.84	1.23
Educating the public about farming	280	3.78	1.11
Educating/teaching new generation to farm	278	3.73	1.3
Reducing carbon footprint	279	3.68	1.15
Maintaining tradition of farming in family or in the community	279	3.63	1.3
Growing consumer demand for organic	278	3.56	1.25
Getting a premium for organic	276	3.26	1.35
Accessing new niche markets (e.g. local, carbon-free)	275	3.25	1.42

(Maine Organic Farmers and Gardeners Association); (2) Community Supported Agriculture (CSA) and local markets; (3) customers and consumers; (4) other grants, organizations, and workshops; (5) off-farm income; and (6) family.

The major themes that emerged from responses to, "What constraints, if any, prevent you from implementing farm practices that align with your priorities as rated here?" (see Table 5.3) were: (1) money, costs, and finances; (2) government and regulations (both federal and local); (3) capital and costs of inputs (labor, seeds, irrigation, soil); (4) getting a premium in the market and consumer knowledge of the product; (5) lack of infrastructure and inability to secure grant funding; and (6) time.

Phenomenological findings

Figure 5.2 illustrates key themes that emerged during the farm visits and interviews organized into the categories of Justice, Logistics, Rationale for Organic Farming, and Networks/Coalitions. Farmers and workers alike had a great deal to say about their conceptualizations of justice and how it is or is not realized in their experience on the organic farm. In addition, much of what the participants discussed centered on the reasons why they farm organically or work on an organic farm in the first place. From their perspectives, the value that participants realized in living this lifestyle far outweighed the challenges they discussed, which were many and significant, including little to no pay, no health insurance, and the difficulty of the work itself. In what follows, we present data that speak to the values-based motivation to farm organically and/or work on organic farms.

Rationale for organic farming

Both the farmers and workers alike discussed their rationale for participation in organic (rather than conventional) farming. Each individual has his/her own reasons for doing so, but they centered on what organic farming symbolized to them idealistically, the sense of community they feel with other organic farmers and workers, the mission of educating the public, and notable beliefs about the health and nutrition associated with growing and consuming organic goods. In addition, they described their own unique pathways into organic farming, and those who had worked previously in conventional agriculture discussed those experiences, providing further reasoning for getting into organic agriculture specifically.

The ideal of farming

Commonly, participants discussed their ideals related to the notion of organic farming as some of their rationale for involvement with it. Jane (pseudonyms are used for participants' confidentiality) from Farm C captures the sentiments of the farmers and workers interviewed by summarizing, "I'm happy, I don't

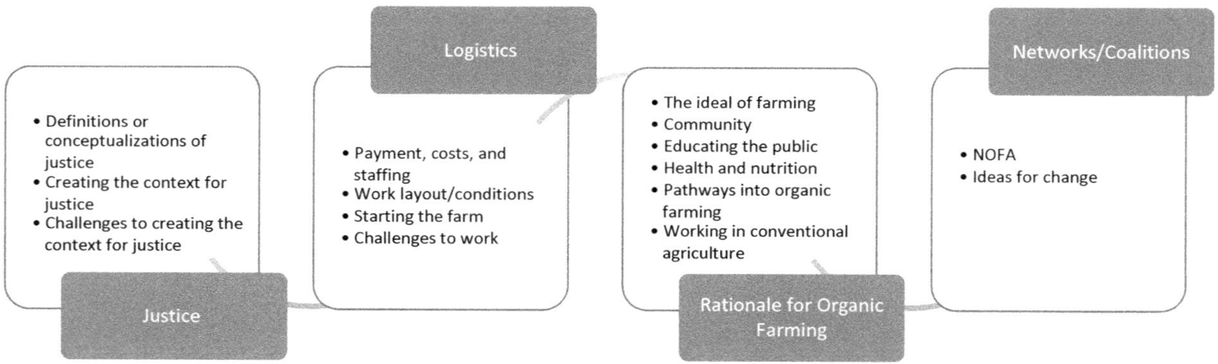

Figure 5.2 Categories and themes from site visits and interviews

need a lot of money. I'm happy doing what I'm doing. I would like to make more money, so I could pay more people." This drive to be happy and to surround oneself with hard workers and community was a motivating factor and ideal of farming for all.

Community

Community was a central tenet to many of the farmers' and workers' values related to organic farming. Whether it is the support of the local community in purchasing the products and volunteering at the farm, or the concept of a community of farmers and workers alike to help support organic farming efforts more broadly, the importance of community was mentioned throughout the interviews. Thinking back on her many years at Farm A, Jen states, "When you have small communities where people watch out for one another, both kicking people's ass and supporting them positively, you know, that's, those are societies that last."

Seeing the power of the community and the farming itself as a symbol of change was stated time and again as a motivator for when times get tough due to unpredictable weather or a rough growing season, which are inevitable. Repeatedly, the farmers and workers on all three farms stated the sense of community they felt as one of the biggest benefits of their work. Ashley, a worker on Farm B summarizes this by saying, "To accomplish all of your goals, which is why you have community and friends and family to help support you when you're, like, unable to make it yourself."

Educating the public

Education was a major focus area for all three farms, centering on different areas, such as: educating potential customers about their goods and services, as well as the value of buying local and organic; working with children and historically marginalized populations to learn about farming itself; educating their workers about organic farming practices; and exploring the connection between their products and health. Their commitment to working with people from all walks of life in understanding the worth of organic farming and its contribution to personal and community health was clear across all three sites.

Health and nutrition

Holistic, natural health and nutrition, not surprisingly, were major foci for most of the interviewees, farmers and workers alike, whether it was a focus that originally attracted them to organic farming or a by-product of the lifestyle they live while on-farm. Joe, a worker at Farm C, talks about health and nutrition as a big motivating factor for getting into organic farming:

> You know, for myself, for going on a few years now, maybe, like, the past, like, five or six years, sort of been developing, like, paying more

> attention to what I eat. And also with that, you know, paying attention to medicine and using, like, herbs. And over the course of that time, I think that I sort of started to share my information with other people to the point where people actually started asking me for help with things. And so then I decided I wanted to just to study it.

As they learned more about health and nutrition, many interviewees also wanted to share that knowledge with others. However, Jane from Farm C points out a particular struggle with messaging the health benefits of their herb business, stating:

> We're like, "No, herbs are medicinal. No, we're not selling medical marijuana, this is medicinal herbs," you know, so. But at this point we, I definitely have developed a reputation in Rhode Island and we work with different doctors and acupuncturists now, and they send people to us for advice when they're kind of stumped.

Pathways into organic farming

All participants discussed how they became interested in organic farming. It is noteworthy that many of the participants had attended college, and that for many of them it was during this time that they became interested with the ideals that ultimately led them to organic farming, sometimes influencing them to leave school to pursue their passion. Will from Farm B left college for a while to pursue farming, but ultimately returned to school to finish his degree:

> So I went back. Decided to go back to school. I studied sustainable agriculture. And really focused on it. And I went from, you know, like, a kind of a B, A minus student to just a straight A student. I was, like, "I'm motivated. I know exactly what I want to do."

Jane from Farm C took a different path into farming:

> I started volunteering at a herb shop in 1999, and then I ended up working there for three years. Yep, in Providence. And then I ended up working there for three years. And then I got a job on a farm for four years. And then, I started just going to different schools in the northeast.

Mike from Farm B states the influence of his college education on his interest in food justice, "Yeah, I think that started when I came up to Albany. I went to school in Albany. So I started, I was studying economics with concentration in urban studies and planning."

Many of the participants were also involved with activist pursuits before or during their time working on organic farms. Bob from Farm A illustrates this, stating he "did anti-war organizing stuff like that, but also did some official community organizing on housing issues." Some participants became involved with organic farming because they wanted to learn how to grow their own food. Mike from Farm B explains, "But I had shopped local. Been a farmer's market kid for a while. And so there was kind of the, I want to learn to grow my own food." Dave, the worker on Farm A, was born abroad and he made his way into organic farming through his post-incarceration program. Only one of the participants discussed being raised on a farm, which is interesting to note but perhaps not surprising due to the small number of individuals interviewed; Jen, the farmer from Farm A, was raised on a conventional farm in Illinois.

Justice: definitions and conceptualizations, creating the context

Several themes emerged that centered on justice. Workers and farmers were asked explicitly about their understanding of justice as it relates to organic agriculture, so some of the excerpts directly reflect participants' answers to that line of questioning. However, sometimes participants' perceptions and understanding of justice as a concept and how it is realized in practice, including challenges and supports for creating a context for justice, were more implicit in nature; therefore, findings in this category center around definitions and conceptualizations of justice, creating the context for justice on organic farms, and the specific challenges that arise in relation to creating that context, based on participants' understanding of justice as a concept.

Both farmers and workers were asked about their understanding of justice in relation to organic farming. Overall, themes such as access, equality, involvement, and education emerged repeatedly as organizing considerations in how these farmers and workers understand justice. For example, Beth at Farm C describes justice as:

> I think the most good for the most people if possible. When I think of food justice, I think of not compromising on any level like respect and good conditions and wages for the worker, and the farmer. If it's a meat farm, good conditions for the animal and then not just people with the money being able to eat it too.

Most participants characterized justice as equality, and not just for all human participants in the system, but for the environment and animals, too. In addition, the concept of personal agency, or having the means to live life the way one sees fit, also emerged as a key tenet of justice.

Many times, farmers and workers talked about how they try to either provide a just working environment or how they are participating in one. Jen, one of the farmers from Farm A, stated, "And a couple of the guys once they graduated from the program, we really wanted to hire them, because they were great, and they needed work. And, so, we kind of ramped up the farm to pay the ex-cons," illustrating their commitment to continuing to provide employment for people marginalized within society. She goes on to say, "I always try to find something for people to be in charge of, whoever they are, whatever they're good at," showing how they play to people's strengths and assets at their farm.

In another example, Beth, a worker at Farm C, notes:

> I think that just knowing that I'm not working with pesticides that means that I feel better, I have better conditions. I think that since it's organic farms I'm working on, usually they're pretty small, and they're not paying people by the piece, they're paying by the hour.

This statement illustrates the relationship between organic farming, payment, and what Beth considers just treatment. Some of the participants went so far as to recognize that creating the context for justice is not entirely an economic concern, but rooted in cultural and human rights as well, allowing laborers the possibility of learning about their own ancestral history with the land and pathways to working toward their own holistic health.

Alternatively, both farmers and workers discussed the different challenges to living out their philosophical values focused on justice. Most of these are logistical in nature, relating to the pressures of running a farm, payment, and housing, and keeping up with ever-changing policy. Bob, one of the farmers at Farm A, points out:

> And plus we got workers' compensation, so we've got other holdings and things like that. At some point, you know, it doesn't make enough to pay for that all the time. And if these new safety laws come in, it's gonna be very difficult to run farms.

Difficulty in compliance is mainly an economic concern, both having the money required to pay into different funds as necessary, but also having the capital (human or economic) to ensure that the farm is upholding its accountability. Processing paperwork, interpreting case law, and figuring out the legal commitments of the farm are things that take time and specialized understanding.

Joe, a worker at Farm C, describes the challenges with finding suitable employment:

> I think employment is a big issue. You know, just being able to kind of sustain employment. You know, for one thing, it's mainly seasonal work. So, there's this kind of, I see a lot of people who work on these farms. Kind of like every season, they're trying to find some new work and trying to find enough work to be able to make ends meet.

The seasonality of the work emerged repeatedly as a challenge for workers, and was recognized by farmers as something that keeps them from employing people full-time, year-round on their farms. All of the participants indicated a lack of appropriate health insurance and other benefits. At Farm A and Farm B, the farmers had off-farm employment to supplement their farm income. Will, the farmer at Farm B, notes that once the farm makes more money he hopes to,

"Start to pay myself a salary. And then the goal is in two seasons to have a regular salary for myself and one other person. So that's kind of, I don't know if that's a dream."

Another common challenge to creating the context for justice is building a consumer base and balancing the need to make a profit to cover expenses with the desire to serve underrepresented populations. Another tension arose around what some of the farmers mentioned as the "entitlement" of some of their customers, who expect fresh produce that looks like it came from a store and how that is at odds with their outreach to people in lower-income populations to make their products more affordable.

Summary

In summary, the survey findings and interviews indicate that most farms in the NOFA network are small-scale organic farms, a population on which little specific research has been done. When looking at the themes that emerge from the survey findings about the supports for and constraints around implementing their values as ranked, they were largely consistent with data from the interviews. All data from both phases of the study illustrate the understanding and desire on behalf of the farmers to provide a just working environment, and the tensions encountered (financial, time, policy-related, etc.) while trying to create that environment on individual farms within the larger political economic landscape of small-scale organic agriculture, which largely disadvantages the farmers and workers, despite a larger public narrative that expresses a desire for more locally focused and just food production.

Justice claims (recognition, capabilities, and participation)

Based on farmers' survey responses, it is clear that the main motivation for many organic farmers in the NOFA network relates to environmental sustainability and an understanding of the health benefits of organic production, and has less to do with drawing a premium within the market. Open-ended survey responses identified a number of the challenges faced by organic farmers in the Northeast, chiefly related to lack of funding and profit, lack of experienced and willing workers, and lack of time to spend both farming as well as completing the necessary governmental and other paperwork needed to flourish. All of this has important implications for framing this population within the theories offered by environmental justice, as these are people who, in many cases, choose to put themselves into a position of marginalization because of the alignment of their values with their lifestyle and practices, despite the multiple hardships they and their workers face. This illustrates how committed some people are to transitioning the U.S. agricultural system, but also highlights the need for change to occur beyond the level of individual farms.

The findings showcase the necessity of including the historical perspective and broader socio-economic considerations, as Callewaert (2002) suggests, in analyses of and organizing for environmental justice. Because the system within which small-scale organic agriculture in the Northeast exists is flawed and focused on growth within a capitalistic economy, the true value of the work and the intention of the farmer and the workers are not typically realized, as noted by both the farmer and worker participants, as well as the members of the Interstate Council. On every farm, the farmers viewed the workers as a part of their family, symbolized in small ways like sharing meals together and in large ways like on Farm B, where their apprentices live in the home with their family. In addition, on all three farms, the farmers were clear in their messages about the values and focus for their farms and their work, and the workers not only understood these values, but also felt deeply and personally connected to them. Education on organic farming and their specific products was a central tenet on each of the three farms, in addition to structuring positions around the workers' skills and interests. On each farm, both the farmers and the workers talked about balancing the necessary farm work with creating opportunities for continual individual and organizational growth. This was seen in tailoring positions to everyone's strengths at Farm A, the apprentice-driven construction of a greenhouse at Farm B, and the opportunity to work in exchange for herbal education at Farm C.

Finally, the interview data reflect the need for farmers to have support, such as educational opportunities, a collaborative market community, and policy updates and assistance in place. These supports and others can allow farmers to truly live and work in alignment with their values and beliefs and to recruit workers who are committed to similar values in order to create the most fruitful environments possible for both parties on organic farms in the Northeast. Findings from the interviews with the Interstate Council are consistent with these themes, but also diverge in that they focus more on the challenges facing the network, the variety of motivating factors that bring people to the table, and the role that the organization can play in shaping the broader system within which organic agriculture exists through policy change, certification, and other measures.

Pellow and Brulle suggest that we must "take seriously the obligation to propose new directions for society to heal itself and produce more just and sustainable forms of production" (2005, p. 296). They also argue that it is crucial to discover at what scale advocacy, participation, and change within the system are most possible. The findings indicate that possibility for change exists at a regional scale, as demonstrated by the participants' recognition of NOFA as a resource, as well as their own voices in shaping the organization itself as well. Farmers in this study saw the opportunities and strength that their regionally networked system created, and many named their own agency in creating a world in which they wanted to live by structuring their life and work around their strengths, capabilities, and interests.

Walker (2012, p. 62) noted that "Qualitative methods have also been used to reveal the values and subjectivities that underpin how different environmental goods and bads are understood by different social groups." In this study, data emerged in alignment with contextual process claims, particularly those that addressed current policy and conditions. However, data more so reflected a shift toward structural process claims that provide insight into the systemic structures within which small-scale organic agriculture in the Northeast exists. Values and subjective perspectives on the process and worth of organic farming framed much of what the participants discussed. Particularly in the site visits, many interviewees reframed things that would typically be understood as inherently bad, as something they were willing to deal with to do work they believed in, reinforcing their commitment to creating structural processes with which their values align. They were willing to live without steady income and/or benefits, for example, to know that they were feeding their families and communities healthy, nutritious food. Across the interviews and survey, participants offered their perspectives on their understanding of justice as it relates to organic agriculture. Results ranged from perceiving justice as equitable treatment to conceptualizing it more broadly as the legitimate involvement of all stakeholders in the different processes affecting them, which showcases a broader conceptualization than solely traditional notions of distributional justice.

It is clear across the survey and interview data that farmers are not in small-scale organic farming for the money, but primarily because their work is rooted in what they value. They do value the economic viability needed to continue their work, but also place great value on producing high quality and nutritious products, avoiding chemicals for the health of the environment and humans, on a strong sense of ecological stewardship, and maintaining ethical trade relationships. Participants also indicated valuing educating the public and consumers about organic practices, providing an equitable and just working environment, and sharing potential health and safety issues with their workers and their communities. Farms chosen for the site visits placed an even higher value on education, equity, and involvement than did the survey participants due to the sampling method used to identify and select them; nonetheless these values showed up across many participants.

Themes emerged across data sources that identified similar supports and constraints in achieving justice. Most of the supports centered on the community of the farm itself, the family and/or members supporting it, the alignment with local consumers who recognized the value of organic agriculture, the network support offered from NOFA chapters, and ongoing educational opportunities about organic practices that help the business aspect of the farm run more smoothly, such as grants and other financial supports. The constraints or challenges in creating the context for justice were also consistent. These revolved around navigating the governmental bureaucracy surrounding organic agriculture and farming practices, the sheer cost of operation while lacking a venue for getting a premium price for goods on the market, time, the wherewithal to navigate alternative sources of funding such as grants, and finally an inability to retain and sustain a vibrant, educated, and passionate workforce over time.

Obviously, many of these constraints are logistical in nature and intertwined with the political-economic context in which organic farms operate, indicating that issues in creating the context for justice are not about a lack of understanding or commitment to justice on the farmers' part, but rather are external factors over which farmers individually have little or no control. Because the farms are small-scale and many times are not as economically viable as they would like to be, in addition to the challenge of finding experienced workers and the seasonality of the work, it can be difficult to retain workers. Also, while salary and benefits are important, an increase in these alone does not equate to a more just or equitable working environment. Other factors such as a sense of community, value placed on working the land, and other contextual factors play a large role in justice for farmers and workers as well, as illustrated during the interviews. The participants knowingly placed themselves in a marginalized role because the benefit outweighed the cost; their belief in the value of farming and laboring organically trumped the challenges they faced.

More research is needed regarding the threshold at which farmers are 'giving up' on justice. While the farmers indicated in the survey and interviews the desire to create a just, equitable working environment, it became clear that other pressures often keep this from being top priority or being fully realized. Again, when farmers are barely able to make a living or provide benefits for themselves and their families, how possible is it to create just conditions for their workers? A better understanding of the bare minimum that farmers need to retain a focus on justice would be useful in focusing the work of organizations like NOFA on helping farmers reach that point.

Disagreement exists about the value and impact justice certification would have in the market(s) in which these small-scale organic farms participate, as well as about the benefits and potential complications such certification would create. Justice certification involves certification based on ensuring fair treatment of workers, fair pricing for farmers, and fair business practices (Agricultural Justice Project, n.d.). While the intention is to create a more just working environment for laborers, the impact could be potentially harmful to farmers already strapped for resources or struggling to make ends meet. Thus, this area needs further research so that informed decisions can be made about participation in such certification initiatives.

Finally, while this study showcased assets present across this broad, regional network, it did not discuss how leveraging these assets within the organization influences farmers' and workers' perceptions over time. In addition, while it uncovered the need for mechanisms facilitating participation across such large, decentralized networks, it did not examine specific strategies for doing so. More research in this area can aid practitioners in asset-based coalition building focused on justice.

How should things be?

Most farms in the NOFA network are small-scale organic farms, a population on which little specific labor-related research has been done. Findings indicate that these farms rely heavily on their families and communities to operate, and that the farmers have a commitment to farming organically, despite the challenges they face financially and physically, because it aligns with their values and commitment to healthy, nutritious food and more just, equitable communities. The biggest challenges faced by these farmers are financial and having the time and infrastructure necessary to navigate policy and develop markets within which their goods can earn a premium. Additionally, the lack of skilled, trained workers and the means to keep them employed due to the seasonality of the work and the challenges mentioned previously are huge hurdles these farmers face. Among the workers interviewed, the biggest challenges faced are the lack of year-round employment, issues with transportation and housing, and the lack of benefits and pay. Like the farmers, workers stay on the farms despite these challenges, because of a commitment to the ideals of organic farming or to learn the skills necessary to farm organically themselves.

While these structural challenges exist, the farmers and workers bring numerous assets to the NOFA network, and the value placed on organic farming outweighs the costs for the farmers who participated in both the survey and interviews. Opportunity exists to leverage these assets, including knowledge within the network on best practices related to worker justice, and to showcase possibilities for aligning practices with values without forsaking the financial gain necessary for farmers to pay themselves and their workers. In addition, building upon these assets at a scale broader than that of the individual farms within the network to influence systemic change for environmental justice in U.S. agriculture, in areas such as policy and certification, has the potential to build upon farmers' and laborers' stated values to create avenues for them to live more in accordance with those values.

Notes

1 Much of what is included in this chapter is sourced from my dissertation research at Antioch University New England, which can be found here: https://etd.ohiolink.edu/ap/10?216504457693527::NO:10:P10_ETD_SUBID:98613

2 The survey was administered online from January 2 to March 15, 2013. An invitation to participate was sent electronically to all members (approximately 1,500) in NOFA through their chapters in seven Northeast states (Connecticut, Massachusetts, New Hampshire, New Jersey, New York, Rhode Island, and Vermont), using a variety of listservs that reach NOFA farmer members. The link to the survey was also distributed through various email newsletters and word-of-mouth at conferences and meetings. While it would be desirable to generalize the survey results to all of the farmer-members of NOFA and/or organic farmers in the Northeast, it is important to note that the data collected are only representative of those farmers who completed the survey. This is because the sample frame, or the list from which the sample was drawn to represent the survey population, was unavailable. Thus, the sample consisted of all NOFA members and organic farmers who received an invitation to participate and then chose to complete the survey, consistent with a volunteer sampling method. While all members of each NOFA state chapter had access to receiving the survey through email distribution and information at their annual meetings, the results of the survey are not representative of the whole population, but rather describe the opinions and experiences of those who completed it (Dillman et al., 2009, pp. 41–64).

3 Interviews lasted from 20–90 minutes depending on the interviewee's responses and follow-up questions necessary for clarification. Questions for farmers focused on their biggest challenges and supports, their understanding and conceptualization of justice, and what they derive from membership in NOFA and other networks in which they participate. For workers, the interview questions also focused on their biggest challenges and supports, understanding of justice, how they got into farm work, and other types of employment they may currently hold. Audio recordings of the interviews were transcribed for electronic storage and analysis.
4 Interview data analysis: Data were coded and categorized into themes with the aid of the qualitative data analysis software Dedoose (www.dedoose.com). Qualitative interview data were analyzed in an ongoing, iterative cycle through four steps: (1) gather data; (2) examine the data; (3) compare the new data to prior data; and (4) identify new data to gather (Lapan & Quartaroli, 2009). Interview data were coded using a 'bottom-up' or 'open' approach, meaning that codes and labels emerged from the text of the interviews itself. Codes were checked throughout both the initial and focused coding processes to ensure they were "being consistently used to mark the same concepts, behaviors, or values" (ibid., p. 266) and patterns were identified by looking for repeated occurrences, and condensing where appropriate, and by the recognition of "what is missing that might be expected to be present; this is a pattern resulting from omission" (ibid., p. 268). I was able to spend informal time with each participant in addition to the formal interview, which allowed me to gauge participant reactivity to me as the researcher. In all cases, participants were consistent in both the informal and formal settings, and seemed therefore to feel free to share their honest opinions with me.
5 It is important to note that each of the three participants has been involved with NOFA for some time (a minimum of ten years), one since 1983. Two of the participants are farmers themselves at some level (one recreationally, the other as a main source of income), and the other left academia to work for NOFA directly. Note that while two of the farmers interviewed on the site visits are members of the Interstate Council, their interview data was not included here because the line of questioning focused on their work as it relates to farming specifically.

References

Agricultural Justice Project. (n.d.) Food Justice Certified. Available at: http://agriculturaljusticeproject.org/?page_id=13 (accessed July 19, 2015).

Arcury, T. A., Wiggins, M. F., & Quandt, S. A. (2009). Conclusions: An Agenda for Farmworker Social Justice in the Eastern United States. In S. A. Quandt, & T. A. Arcury (Eds.), *Latino farmworkers in the Eastern United States* (pp. 221–233). New York: Springer. Available at: www.springerlink.com/content/k4138470 31p03243/abstract/

Berkey, B. (2014). Just Farming: An Environmental Justice Perspective on the Capacity of Grassroots Organizations to Support the Rights of Organic Farmers and Laborers. Keene, NH: Antioch University. Available at: https://etd.ohiolink.edu/ap/10?216504457693527::NO:10:P10_ETD_SUBID:98613

Berkey, B., & Schusler, T. (2016). Justice Issues Facing Family-Scale Farmers and their Laborers in the Northeastern United States. *Journal of Agriculture, Food Systems, and Community Development*, Advance Online Publication. Available at: http://dx.doi.org/10.5304/jafscd.2016.062.017

Bryant, B. (1995). *Environmental justice: issues, policies, and solutions.* Washington, DC: Island Press.

Callewaert, J. (2002). The Importance of Local History for Understanding and Addressing Environmental Injustice. *Local Environment*, 7(3), 257–267. Available at: http://doi.org/10.1080/1354983022000001633

Creswell, J. (2009). *Research design: qualitative, quantitative, and mixed methods approaches* (3rd ed.). Thousand Oaks, CA: Sage Publications.

Dillman, D. A., Smyth, J. D., & Christian, L. M. (2009). *Internet, mail, and mixed-mode surveys: the tailored design method*. Hoboken, NJ: Wiley & Sons.

IFOAM (International Federation of Organic Agriculture Movements). (n.d.a). Leading, Uniting and Assisting the Organic Movement since 1972. Available at: www.ifoam.org

IFOAM (International Federation of Organic Agriculture Movements). (n.d.b). The Full Diversity of Organic Agriculture: What We Call Organic. Available at: http://infohub.ifoam.org/sites/default/files/page/files/full-diversity-organic-agriculture-en.pdf

Lapan, S., & Quartaroli, M. (2009). *Research essentials: an introduction to designs and practices* (1st ed.). San Francisco, CA: Jossey-Bass.

Locke, L. F., Spirduso, W. W., & Silverman, S. J. (2000). *Proposals that work: a guide for planning dissertations and grant proposals*. Thousand Oaks, CA: Sage.

McMillan, J. H., & Schumacher, S. (1997). *Research in education: a conceptual introduction*. New York: Longman.

Morison, J., Hine, R., & Pretty, J. (2005). Survey and Analysis of Labour on Organic Farms in the UK and Republic of Ireland. *International Journal of Agricultural Sustainability*, 3(1), 24–43. Available at: http://doi.org/10.1080/14735903.2005.9684742

NOFA Interstate Council. (n.d.). Who We Are. Available at: www.nofa.org/whoweare.php (accessed May 22, 2016).

Patton, M. Q. (2002). *Qualitative research and evaluation methods*. Thousand Oaks, CA: Sage.

Pellow, D., & Brulle, R. J. (2005). *Power, justice, and the environment: a critical appraisal of the environmental justice movement*. Cambridge, MA: MIT Press.

Shreck, A., Getz, C., & Feenstra, G. (2006). Social Sustainability, Farm Labor, and Organic Agriculture: Findings from an Exploratory Analysis. *Agriculture and Human Values*, 23(4), 439–449.

Sze, J., & London, J. K. (2008). Environmental Justice at the Crossroads. *Sociology Compass*, 2(4), 1331–1354. Available at: http://doi.org/10.1111/j.1751-9020.2008.00131.x

USDA (United States Department of Agriculture). (2012, December 5). Agricultural Marketing Service. Organic Labeling. Available at: http://www.ams.usda.gov/AMSv1.0/NOPOrganicLabeling

USDA (United States Department of Agriculture). (2014). National Agricultural Statistics Service. *2012 Census of Agriculture: United States Summary and State Data* (Vol. 1, Part 51). Available at: www.agcensus.usda.gov/Publications/2012/Full_Report/Volume_1,_Chapter_1_US/usv1.pdf

Walker, G. P. (2012). *Environmental justice: concepts, evidence, and politics*. London: Routledge.

6 The current landscape of change

Fighting for justice for agricultural labor

Justice claims: how should things be?

Chapters 2 and 3 covered some of the most pressing issues facing farmworkers today, in addition to the historical and political contexts within which those issues were created and continue to be perpetuated. Many times throughout my own research I have felt paralyzed by the sheer complexity of what it would mean to create positive change within the farmworker community. In addition, I have been struck by what it means to go back in time and look at a system that in many ways remains unchanged despite the enormous amount of energy spent to improve the livelihood of farmworkers on behalf of many farmworker advocates over the recent decades.

As long as we have a food system in which people expect to get what they want at the prices they want to pay (read: cheap food), the bottom of the chain of food production (read: labor) and those in it will continue to be squeezed, their rights subsumed by every other part of the chain so that our local fast food chains can continue to offer dollar menus and supermarket corporations can continue to profit. As Melinda Wiggins (2009) points out, the current agricultural industry relies on a disenfranchised and easily exploitable workforce who cannot control the conditions of their own work, due to the agricultural exceptionalism discussed in Chapter 3. In addition, the general public's notion of the array of factors impacting farm labor is often skewed, equating labels like organic with better outcomes along the chain. As we saw in Chapter 5, labor within organic agriculture, particularly small- and family-scale organic agriculture, is complicated by its place within the larger political economic system of agriculture. The organic label in and of itself is no guarantee about anything regarding labor, including better treatment and/or the ability to make a living. As long as the costs associated with our eating priorities continue to be externalized, those who labor on our farms and in other parts of our food system will continue to face injustice.

Chapters 4 and 5 highlighted in depth the work being done to positively influence the conditions of farmworkers, both by an organization focused on farmworker rights specifically and by a group with a broader focus concerning farm labor. Even with these examples, agriculture in the United States is a wicked system, and it is hard to imagine what positive change could look like. Justice for farmworkers is further complicated by the fact that the majority of those laboring on America's farms are not legal citizens of the United States, and as migrant workers are a largely mobile population: hard to pin down, difficult to study and understand for the purpose of policy change, enforcement, and advocacy. Then there is the question of what type of change is needed and which areas to prioritize, and the fact that change itself is inherently difficult in the American political system. However, not all is lost.

More people than ever before do care about food: where it comes from, under what conditions it was grown and cultivated, its potential impact on our environment and health. Given this fact, there is an opportunity to bring farm labor to the forefront and to involve the consumer base, policy-makers, and every person who has a stake in our food system (hint: that is everyone) in the conversation about what happens to the *people* who cultivate our food. There is a movement that has deep roots around farmworker rights both in dignifying the work of farm labor, and in advocating for legislation that can positively impact farmworker livelihoods and health. All of this must take into account what John Green, Eleanor Green, and Anna Kleiner assert in saying that there is, "limited knowledge of, participation in, and access to government agricultural programs, and the existence of few programs specifically designed to assist small-scale and limited-resource producers" (Green et al., 2011, p. 56). All of this is to say that while farmworker rights are being considered, the larger system itself neglects to bring its protections to the forefront of farmworkers' lived reality, and also seeks to actively support the large-scale agriculture within which farmworkers are often rendered invisible.

In this chapter, I will review some of the literature that upholds justice claims for farmworkers, posing the question: "How should things be?" and advocating for models of coalition building, community development, and social change that can inform the broader movement as it considers recognition, capabilities, and participation of farm laborers in the very systems that impact their lives. In addition, I will highlight the work of selected movements, organizations, and coalitions to give examples of how these models play out on the ground. Finally, I will conclude this chapter with a brief exploration of fair trade labeling and its reflections of and implications for labor.

Environmental justice, coalition building, and community change: a review

As reviewed in Chapter 1, the notion of social capital is about the connections between and among people, as well as our capacity to use relationships as a resource to accomplish and work toward our goals (Phillips & Pittman, 2009). It has become clear through our exploration in the first few chapters of this book that building social capital among and with farmworkers would have far-reaching implications. They currently lack the social capital and capacity for strategy enactment. There are, however, approaches to building this. Schlosberg states that the field of environmental justice demands "authentic, community-based participation [that] comes out of the experience of disenfranchisement" (2007, p. 68). Environmental justice from this vantage point and with authentic stakeholder participation, when applied to rights for immigrant farmworkers, holds untold possibilities for advancing the cause of justice for those who toil in our fields. In this section, I will explore key elements of the seminal literature of power and privilege, coalition-building and community change, and advocacy research to provide a foundational framework to achieve an expansion of environmental justice that can help the plight of farmworkers in U.S. agriculture.

Power and privilege

Critical theory involves analyzing and critiquing present oppression while simultaneously exploring an emancipatory or reconstructive goal that allows the oppressed to move beyond their current conditions. Therein, by discovering the context for one's own oppression and exploring it, one can begin to break away from it. As Paulo Freire points out:

> Domination, by its very nature, requires only a dominant pole and a dominated pole in antithetical contradiction; revolutionary liberation, which attempts to resolve this contradiction, implies the existence of not only these poles but also of a leadership group which emerges during this attempt.
>
> (2000, p. 132)

He goes on to say that this leadership necessarily must identify with the oppressed group, thinking *with* this group rather than *about* it. Robert Chambers discusses uppers (dominators) and lowers (those dominated) using powerful language:

> The upper rejects discordant feedback. He re-labels the lower and redefines her reality for her. He seeks to transfer his reality. He wants it to be his reality that counts. What prevents him from learning in this case is not professionalism or bureaucracy. It is his dominant behavior, person to person. It is his power.
>
> (1997, p. 75)

In all societies, there are people who have more than others; more money, more influence, more opportunity, and so on. In some societies, such as in the United States, the distance between those who have and those who do not is immense, setting up the dynamic for exploitation. Johnson (2006) explains two types of privilege: unearned and conferred. The most surprising part of the dominance that power and privilege allow one to assert, based on his/her group identity is that it seems normal, based simply on the way society is structured, that it would be distributed unequally. People do not question, for example, a man's qualifications to be in a position of leadership the same way they question a woman's, due to the patriarchal set-up of societal hierarchy. This phenomenon is commonly referred to as cultural hegemony, but seldom in our media and communal dialogue do we hear people questioning the "way things are," usually because we are so immersed in the world the way it is to even recognize the tenuous dynamics at play. As Freire puts it:

> The oppressor elaborates his theory of action without the people, for he stands against them. Nor can the people – as long as they are crushed and oppressed, internalizing the image of their oppressor – construct by themselves the theory of their liberating action.
>
> (2000, p. 183)

When it comes to agricultural labor in the United States, many are all too aware of their place in the societal structure, and these power dynamics have frightening consequences. As we have seen, oftentimes these farmworkers are mistreated by overexposure to pesticides, wage theft, unhealthy living conditions, and other injustices, but they do not feel empowered to report these human rights violations. Particularly for immigrants, due to threats of deportation and harm to their families in their home countries, these farmworkers become victims of their own realities. Those with power and privilege (corporations, farmers, U.S. citizens, etc.) then become content to find fault in the immigrant farmworker for putting him/herself into a position of abuse, rather than blaming the abuser or the failure of the system itself. Chambers describes:

> To blame the victim is perhaps the most widespread and popular defense. It flatters the uppers and flattens the lowers. There is a hierarchy of acceptability to uppers of different explanations of error and failure. Uppers usually prefer to blame lowers, and usually prefer explanations and prescriptions which require that lowers change rather than that they themselves or other uppers change.
>
> (1997, p. 83)

Chambers explains that this transfer of reality guides action and research, and can have dangerous implications when one works with vulnerable populations. It is not enough to simply understand the 'other' in working with marginalized groups, one must also be willing to make his/her own reality count less, so that of the disadvantaged can count more.

This intersects with the critiques of the environmental justice movement as traditionally realized and practiced, discussed in Chapter 1. Many environmental justice tales consist of documenting the woes of the disadvantaged at the local level, rather than on truly assessing what needs to be changed on a grander scale. A solution is often sought to simply solve the immediate issue, instead of also working toward progress on the dynamics underlying it. For example, it is one thing to examine the impacts of pollution on a community, but it is another entirely to involve those whom the pollution is affecting in examining it *and* the political-socioeconomic conditions that caused it in the first place. The second threatens the power of the privileged by exposing the hegemonic structures within which the injustice occurs, and empowers the disadvantaged to take control into their own hands. This addresses the dimensions of participation, capabilities, and recognition that Schlosberg (2007) encourages and expands the possibility of environmental justice to truly shift and change the power dynamic.

This fundamental shift addresses a "very obvious truth: just as the oppressor, in order to oppress, needs a theory of oppressive action, so the oppressed, in order to become free, also need a theory of action" (Freire, 2000, p. 183). Inquiry should aim to humanize those marginalized, and "cannot be carried out in isolation or individualism, but only in fellowship and solidarity" (ibid., p. 85). When we take into account the multiple dimensions of the context for injustices facing those working in U.S. agriculture, we are reminded that:

> The social organization of environmental hazard and the political distribution of risk also manifest strong racial dimensions. The environmental racism literature provides convincing evidence of disproportionate exposure of poorer communities of color to the most hazardous forms of environmental pollution.
>
> (Newell, 2005, p. 75)

To move beyond what is well documented as stated here and to truly make change, we need more than the conventional cataloguing of disaster and exposure. What is needed are strategies to uncover and explore the system itself and to document the relation of these incidences not as unique, but as part of something bigger and more deeply rooted in the dominant socioeconomic system. In doing so, we begin to find grounds for systemic change.

Coalition building and community development

As discussed in Chapter 1, in his book, *Coalitions across the Class Divide*, Fred Rose (2000) offered that synergy between both strategy and circumstance are essential in successful organizing. In addition, he illuminates the fact that effective coalitions are able to recognize the circumstances, opportunities, and limitations impacting how they proceed in positively changing the issue they face. As is predictable, navigating these dynamics becomes more complex as more people, organizations, and communities participate, as they each bring their own set of both distinct and overlapping values, beliefs, and interests. When considering issues facing farmworkers, not only are farmworkers themselves diverse and largely mobile at the individual, group, and community levels; but given the scale of agriculture within which they labor, their experiences vary widely. The circumstances around which to organize exist, but agreeing on them and determining the appropriate strategy is more complex. Also, given that farmworkers are moving constantly in and around this system, we can revisit the question posed previously: how do we create relationships necessary for justice and social change?

Couto (2002) listed values, inclusiveness, initiative, and creativity as central to the democratic leadership needed to build both solidarity and coalitions. Given that there might be varied interests of the multiple stakeholders engaged in changemaking, it is necessary, as he frames, to do several things: 1) agree on a shared set of values, 2) articulate and hold one another accountable to strategies for inclusiveness, relying on one another, 3) create ways for members to take initiative when they have a strength, resource, or skill to offer, and 4) foster, rather than suppress, creativity. In doing these things, diverse groups of organizations and individuals can begin to find a common foundation upon which to build strategies for the circumstance they hope to address or advance together.

To do all of this authentically, it is imperative to ensure that the appropriate voices are included in the dialogue. In addition, being thoughtful and transparent about the context for the circumstance is an important consideration as well. Couto (1999) reminds us that when forming relationships and considering strategies, human worth must exceed market and labor values. To summarize his assertions, for coalitions to thrive, they must not simply name the problem, but actively engage in solution-making that acknowledges and involves those least advantaged by the system being changed. As stated earlier, Phillips and Pittman discuss these notions of social capital and capacity as being central to the community development and coalition-building process, stating, "If citizens cannot plan and work together effectively and inclusively, then substantial community progress will be limited" (2009, p. 49).

As previously discussed in Chapter 1, a thorough consideration of the types of public participation is needed when figuring out how to encourage and support authentic involvement from multiple stakeholders. To review, Green and Haines (2001) state that there are four types of public participation: (1) public action; (2) public involvement; (3) electoral participation; and (4) obligatory participation. Giving stakeholders different options for entering the process and participating as they choose is crucial in reaching multiple needs and addressing as many interests as possible. The organizations, coalitions, and movements given as examples in this chapter range in the types of public participation by which they involve farmworkers. Building trust in ways that allows for authentic and unencumbered participation is difficult given the various power dynamics at play, as Ife and Tesoriero (2002) posit. Therefore, expanding participation to include a wider range of perspectives necessitates a period of trust-building between the groups working together, in addition to a clear process for vetting concerns when that trust is broken. Each example given later in this chapter has its own way of addressing this, but in all cases, building trust cannot be ignored. And, given the diversity of these groups and the stakeholders involved, consensus is messy, particularly when that diversity is valued rather than silenced. Consensus, nonetheless, expands the parameters of concern and areas for action. The venues for participation must address this; those involved must be open to the process when there are no easy answers, building upon their trust that everyone has one another's best interests at heart. In addition, these venues must be accessible to all stakeholders, meaning offered at days and times amenable for workers' schedules, in languages that all can understand and contribute using, and other considerations (such as childcare) taken.

Asset-based community development offers a useful framework that connects with critical theory, considerations of power and privilege, and considerations of broader theories of justice. John Kretzmann and John McKnight (1996) coined the term asset-based community development as an alternative path to needs-based community development, with asset-based being focused on building capacity rooted in strength. More than a decade later, Anna Haines (2009) and Rhonda Phillips and Robert Pittman (2009) continue to offer asset-based community development as an alternative to the more traditional needs-based community development, which focuses on the detrimental aspects of a community and the development of strategies for changing it. Kretzmann and McKnight assert that the traditional deficiency-based model of community development is self-defeating in many ways: how can one expect positive development to occur when the areas of focus in and of themselves become the story? If all members of a community, broadly defined for the purposes of farmworkers here, are told to focus on what is wrong within their community rather than their own strengths and assets, the motivation for making change drops dramatically. They state, "Each community boasts a unique combination of assets upon which to build its future. A thorough map of those assets would begin with an inventory of the gifts, skills and capacities of the community's residents," (Kretzmann & McKnight, 1996). These maps involve gifts of individuals, associations, and institutions. All of this aligns with justice claims discussions of capabilities and recognition, illustrating that asset-based community development allows the full perspective of what a community *can* do with its strengths combined.

They also point out two major considerations for asset-based community development: (1) that a focus on assets is not meant to replace a need for additional external resources; and (2) that it is meant to affirm and build upon work already happening around the country. Related to previous discussions of social capital and coalition building, they state:

> If a community development process is to be asset-based and internally focused, then it will be in very important ways "relationship driven." Thus, one of the central challenges for asset-based community developers is to constantly build and rebuild the relationships between and among local residents, local associations, and local institutions.
>
> (ibid.)

A stronger connection is needed with the frameworks offered by power and privilege and their connection to the possibility of building the relationships and trust necessary to unite for change. Examples provided in the next section of the chapter offer some insight into how this might look on the ground.

The fight for social change

Numerous organizations are working for the betterment of farmworkers in U.S. agriculture, using many of the frameworks of social change described in the previous section and in ways that expand environmental justice as called upon in Chapter 1. What follows is a list of a variety of organizations to portray different scales and types of justice. They are groups with which I am most familiar and also those my farmworker activist community connections have recommended as the most representative of the types of work being done. These organizations are coordinated here into three thematic areas, which are not mutually exclusive: (1) farmworker associations; (2) coalitions formed around the justice issue of pesticide use and poisoning; and (3) broad-based organizations focused on farmworker rights. For each organization highlighted, I will include a brief synopsis of their mission and organizational activities, information about how they engage in change making for and with farmworkers, and the scale and location of the organization as applicable. Also useful to consider in reading about these organizations are, as Wiggins (2009) noted, strategies for farmworker advocacy, which include organizing, advocacy, and service.

Farmworker organizations

According to the Food Chain Workers Alliance (Food Chain Workers Alliance, 2012), there are 484 organizations in the United States focused on labor in agriculture, and they divide these into unions (16), worker centers (37), advocacy organizations (28), and service organizations, with the largest number (403) falling in the latter group. At the conclusion of this report, which details the information they found on organizations supporting workers all along the food chain, they state,

> As the food movement continues to grow, food workers organizations are becoming an increasingly important stakeholder in that movement – not only to win justice in the workplace, but to also transform the food system into one that is more sustainable for consumers, communities, and the environment.
>
> (ibid., p. 7)

This report gives the scope of the 1,300+ organizations nationwide that are supporting and involving workers all along the food chain, providing the larger landscape within which the work I discuss in this chapter is placed.

Melinda Wiggins (2009) uses the work of Cesar Chavez and Dolores Huerta in beginning the United Farm Workers (UFW) to characterize the birth of the modern-day farmworker justice movement. In combining "union organizing strategies with civil rights tactics" she states that they "built a base of local workers, which was supported by allies across the country" (ibid., p. 207). She explains how their reach expanded past California, and how "at its height in

1970, the UFW had over 50,000 members and was not only busy negotiating contracts, but was also lobbying for legislative changes at the state level" (ibid., p. 207). According to their website, the "UFW continues to organize in major agricultural industries across the nation. Recent years have witnessed dozens of key UFW union contract victories, among them the largest strawberry, rose, winery and mushroom firms in California and the nation," and "many recent UFW-sponsored laws and regulations aid farm workers; in California, the first state regulation in the U.S. prevents further heat deaths of farm workers. The UFW is also pushing its historic bipartisan and broadly backed AgJOBS immigration reform bill" (UFW, n.d.).

While they are an active organization, the work of UFW has also inspired the approaches and strategies of many farmworker organizations highlighted in this section. In addition, UFW is an example of how politically charged organizations sometimes work well with others to create change. Additionally, UFW exemplifies what Arcury and Marin imply when they state, "These organizations more often work at cross-purposes. Political organizations representing both capital and labor argue that their goals are to improve the agricultural economy while protecting the health and safety of workers" (2009, p. 32). The focus of groups like UFW and others featured in this section on worker health and safety are many times at odds with the agricultural capital focus of other organizations, which can make relationship and trust-building slow, if not virtually impossible. Ultimately, these organizations are change-focused in a political climate that is just as complex as the one that created the justice issues they seek to remedy.

Farmworker associations

The Farmworker Association of Florida

While the Farmworker Association of Florida was the subject of Chapter 4, I reintroduce it here as it uses its foundation in participatory action research to address the forms of power and community building discussed above, builds upon the assets of those involved, provides valuable information, and aims at social change for public good. Greenwood and Levin (1998) and Stringer (1999) discuss action research, and build upon the history of studies in this tradition. They point out as key elements of action research the following: its anti-bureaucratic nature, narrativity, considerations of power, and its existence at the intersection of traditional dichotomies like authority and community, realism and relativism, and epistemology, and local activism. From this perspective "action research lies at the very center of human life" (Greenwood & Levin, 1998, p. 90). Perhaps most unique to this type of research is the position of the researcher, who insists that "research processes, research outcomes, and the application of results to problem solving are inextricably linked" (ibid., p. 93) rather than the traditional social science view of the researcher as an objective party, seeking control. As such, Stringer (1999) points out that action research consists of a different set of principles than other types of social science research. The working principles Stringer puts forth for action research are relationships, communication, participation, and inclusion. The tools offered by action research bear striking resemblance to those of asset-based community development, including asset mapping and working from a place of strength in consideration of community.

Participatory action research builds upon this with an even more explicit consideration of power, involvement, and social change. Fals Borda and Anisur Rahman teach us that, "It should be noted that knowledge itself was not the main motivation for carrying out the research project; rather, the main motivation was to breathe new life into a struggle" (1991, p. 38). The three basic principles highlighted for action and participatory research by Fals Borda and Anisur Rahman are: (1) recognizing the potential for change as well as the positive and recoverable aspects of economies, social lives, and cultures; (2) developing educational activities with the full and active participation of all stakeholders; and (3) creating a horizontal educational process from teaching to learning through a continuing dialogue between professionals and the community. These principles build upon a critical theory framework and are reminiscent of ideas put forth by Freire (2000), and their implications are not straightforwardly a non-traditional role for the researcher but also a need for flexibility in methodology. The 'study' unfolds as the people do. The need for participatory action research is well documented, and "underscored in a large body of research documenting profound differences in health and life changes along racial and ethnic lines and along socioeconomic lines" (Minkler & Wallerstein, 2003, p. 8). The FWAF engages its membership directly in its research projects and partnerships, which have ranged over time from studies on pesticide-related illness, to the impacts of pesticides on women and children farmworkers, to heat stress, and to many other injustices impacting farmworkers.

La Via Campesina

La Via Campesina is an international organization focused on food sovereignty. According to Peter Rosset (2011, p. 22), "Food sovereignty starts with the concept of economic and social human rights, which include the right to food, but it goes further, arguing that there is a corollary right to land and a 'right to produce' for rural peoples." The organization was founded in 1993, and while international in scope, it has an influence on much of the activist work happening in the United States, as seen in Chapter 4 about the Farmworker Association of Florida. According to its website:

> La Via Campesina is the international movement which brings together millions of peasants, small and medium-size farmers, landless people, women farmers, indigenous people, migrants and agricultural workers from around the world. It defends small-scale sustainable agriculture as a way to promote social justice and dignity. It strongly opposes corporate-driven agriculture and transnational companies that are destroying people and nature. La Via Campesina comprises about 164 local and national

> organizations in 73 countries from Africa, Asia, Europe and the Americas. Altogether, it represents about 200 million farmers. It is an autonomous, pluralist and multicultural movement, independent from any political, economic or other type of affiliation.
>
> (La Via Campesina, n.d.a)

When it comes to their work in the United States, they state, "Food justice is one such struggle. The radical roots of food justice in the United States are deep in the movement for Black Liberation" (La Via Campesina, n.d.b), exploring how this movement is caught up in the livelihoods and civil rights-related struggles of racialized peoples in the United States.

The Farmworker Support Committee

CATA (El Comite de Apoyo a Los Trabajadores Agricolas/The Farmworker Support Committee) is a well-known farmworker organization based in southern New Jersey, Pennsylvania, and the Delmarva Peninsula in Maine. The website for the organization states:

> CATA is a migrant farmworker organization that is governed by and comprised of farmworkers who are actively engaged in the struggle for better working and living conditions. CATA's mission is to empower and educate farmworkers through leadership development and capacity building so that they are able to make informed decisions regarding the best course of action for their interests. CATA has advanced based on the belief that only through organizing and collective action can they achieve justice and fullness of life. CATA's programs are based on the Popular Education Methodology, which actively involve farmworkers in the process of social change. This means that the analysis and proposed actions come directly from the farmworkers. Also inherent in CATA's mission is the importance of analyzing the farmworker reality in terms of the food system. In doing so, projects and campaigns are undertaken with the goal of achieving meaningful and lasting improvements rather than mere reforms to a legal and economic system that is structurally biased against them.
>
> (CATA, 2012)

The organization's priorities are migrant and immigrant rights, workers' rights, environmental justice, and housing. In addition to providing programs and organizing with farmworkers, CATA also publishes information and actively networks with organizations across the country and the world to ensure better standards of justice for farmworkers. While they have an extensive partnership list, CATA has been pivotal in the formation of the Agricultural Justice Project and the Domestic Fair Trade Association (covered later in this chapter). They are also active in the United Nations, solidifying the work they do at local, regional, national, and international scales.

In a piece entitled "Race and Ethnicity from the Point of View of Farm Workers in the Food System," Nelson Carrasquillo (2011) reflects upon the growth of agriculture in the United States and its implications on workers. Carrasquillo has worked with CATA since 1992, and he explains that around the time he came to work with the organization, the area's farmworker demographic was shifting from a predominantly Puerto Rican workforce to one comprised of Puerto Ricans and Mexicans. He concludes that because of this shift, "this transition was irreversible and that CATA needed to change accordingly. Our perspective shifted from a Puerto Rican-based organization to a farmworker organization" (ibid., p. 122). Carrasquillo explains this transition towards working with two different populations, overlapping conditions, and diverse ways of experiencing those conditions:

> There is a fundamental difference between the Puerto Rican farm worker and the Mexican worker; first are the issues of citizenship, transportation, and unemployment. In terms of similarities, both groups believed that there was very little they could do to change the working conditions, which in essence were the same for both groups. That spoke to the sense of powerlessness over their working conditions. In both situations, economic necessity was and is their driving force. This demonstrated how those suffering economic and social hardships in their home countries are easily taken advantage of and submitted to unfair working and living conditions. This illustrates how one ethnic group was used against another shows the systematic effort to implement practices that would benefit the farmers.
>
> (ibid., p. 123)

The work of CATA is both on the ground and at a higher level, working with their farmworker constituents for redress, adequate protections, and their enforcement, while maintaining involvement in broader-level farmworker change movements.

Líderes Campesinas

Líderes Campesinas is an organization in California that is focused on women farmworkers' (campesinas') rights. According to its website, its mission is:

> To develop leadership among campesinas so that they serve as agents of political, social and economic change in the farmworker community. This leadership has created an organization by and for campesinas. The approach emphasizes capacity building, democratic decision-making, advocacy, peer training and leadership development as well as a mixture of traditional and innovative education, outreach and mobilizing methods such as house meetings, arts, and theatrical presentations at community venues.
>
> (Lideres Campesinas, n.d.a)

Women who represent the various counties that the organization serves determine its structure and direction; its ultimate goals revolve around the empowerment of farmworker women to be agents in their own lives and leaders in their communities. They use these strategies: "helping and coordinating existing groups of campesinas and those that are about to form; promoting and supporting the development of these groups' leadership skills forming a network of farm-working women" (Lideres Campesinas, n.d.b).

Maylei Blackwell (2006) wrote an ethnography about Líderes Campesinas in which she explains that the organization emerged from another founded in 1988 called Mujeres Mexicanas, which emerged from community-based participatory research being conducted with farmworker women in California. This study led to organizing women in different areas around the state for gatherings in the early 1990s, most notably in 1992 and 1994. Through the study and these gatherings, women identified as their most pressing issues:

> Domestic Violence, Pesticides, Sexual Harassment/Assault in the Workplace, Labor and Hours, Housing, Education, Breast and Cervical Cancer, Childcare, Teen Pregnancy, and HIV/AIDS. Women shared their concerns about labor abuse through managers, lack of communication between management and workers, poor hygiene in the restrooms, scarce fresh drinking water availability, and unfair payment of wages.
>
> (Lideres Campesinas, n.d.a)

As Blackwell states:

> Injury, long hours, heat exposure and unfair labor practices are conditions that campesinas share with their male counterparts but there are some conditions that affect women differently, like not having access to restrooms. Many campesinas develop kidney and liver problems as well as urinary tract infections due to not having access to restrooms or proper sanitary facilities.
>
> (2006, p. 14)

It is for these reasons and many more that Líderes Campesinas exists to provide farmworker women with outlets in their own lives and the tools to organize and be leaders among others in their communities. In another piece, Blackwell (2010) sums all of this up, stating:

> The organization seeks to change the dismal working conditions in the fields and packinghouses and educates campesinas about pesticide exposure, gender discrimination, sexual harassment, and assault. They organize educational meetings where they use teatro and popular education to raise awareness of the issues of domestic violence, the lack of adequate housing and basic transportation for farmworkers, and women's health and well-being.
>
> (2010, p. 19)

Student Action with Farmworkers

Student Action with Farmworkers (SAF), incorporated as a nonprofit in 1992, is based out of Duke University in North Carolina and its geographical area of focus is the southeastern United States. Its mission states:

> To bring students and farmworkers together to learn about each other's lives, share resources and skills, improve conditions for farmworkers, and build diverse coalitions working for social change. SAF works with farmworkers, students, and advocates in the Southeast and nationwide to create a more just agricultural system. Since 1992, we have engaged thousands of students, farmworker youth, and community members in the farmworker movement.
>
> (Student Action with Farmworkers, 2012)

The organization has received numerous awards and accolades, and has accomplished quite a bit in more than twenty years. Among its direct impacts are "80,000 farmworkers and their family members have gained access to health, legal, education, and organizing support provided by SAF interns in the Carolinas" (Student Action with Farmworkers, n.d.). In an article on involving students as researchers in community-based participatory research projects, Pamela Rao, Thomas Arcury, and Sara Quandt state SAF's, "overarching goal is to produce the next generation of farmworker advocates" and that "during their 3-month terms they receive valuable work experience and insight into community work that will benefit them in their careers, as well as some financial assistance for school" (Rao, Arcury & Quandt, 2004, p. 10). Perhaps most compelling in this organization's example is how it has been able to raise awareness of farmworker issues among thousands of people, and provided venues through which they can become active in working toward social justice for workers.

Coalitions formed around the justice issue of pesticide use and poisoning

Coming Clean: Farmworker Health and Justice Team

Coming Clean, Inc. grew out of the antitoxics branch of the environmental justice movement and is an organization with multiple working groups that are all aimed at the reduction of toxic chemicals in everything from the workplace to products used each day. They acknowledge the adverse effects of petrochemical use in our everyday lives, and most directly as they manifest in declining health among individuals and communities. In working toward these goals, they state:

> A new strategy and structure were needed to enable groups to collectively focus on national reforms in a manner that simultaneously increased the

> strength and reach of their own organizations in connection with other groups nationwide. Coming Clean was founded to provide a collaborative intelligence building and campaigning space for such groups.
>
> (Coming Clean, Inc., n.d.a)

Among their ten working groups is one focused exclusively on farmworker health and justice. This working group has representatives from multiple organizations, including many of those featured in this chapter. Using their collaboration in this space, this working group is active in a number of initiatives pertaining to farmworkers' health and exposure to pesticides. They were instrumental in the campaign to update the Worker Protection Standard (discussed in Chapter 3). As a group, they focus on how to ensure that farmworker voices are included in conversations about pesticide-related policies that impact their lives. They do this by directly connecting farmworkers to policymakers, and by working to ensure distribution of farmworker stories and media coverage of farmworker pesticide issues, "helping to get farmworker stories told and voices heard on the national stage" (Coming Clean, Inc., n.d.b).

Toxic Free North Carolina

The mission of Toxic Free North Carolina is to, "engage North Carolinians in the transition to a toxic-free society through initiatives that promote human and environmental health" (Toxic Free NC, n.d.). The organization does this by educating people about pesticide hazards and alternatives to pesticide use, holding government organizations that oversee the use of pesticides accountable for enforcing policies, and by helping those marginalized to see their role in making positive change in the reduction and elimination of pesticide use. This organization offers a number of volunteer and educational opportunities, and provides a useful example of a statewide network aimed at making change and informing people of the current situation regarding pesticides and their effect on us all. Among its central values are universality of human rights, sustainability, accountability, accessibility, and vitality. While not explicitly a farmworker organization, Toxic Free North Carolina recognizes the impacts of pesticide use on the farmworker community and is active in networks seeking to reduce farmworker exposure, strengthen policy and enforcement, and educate farmworkers on their rights.

The Farmworker Health and Safety Institute

The Farmworker Health and Safety Institute was formed in 1992 as a coalition between FWAF and CATA, along with another organization called the Border Agricultural Workers Project (BAWP), formed in 1992 (BAWP became a member in 1996). According to their website:

> The Institute has conducted its innovative training programs with a focus on capacity building through teaching farmworkers to organize around environmental issues and how to impact public policy. As a result of sharing these experiences and technical expertise with farmworkers in the United States, Mexico, Central America, and the Caribbean, the Institute has evolved into the training and monitoring arm of its member farmworker organizations.
>
> (Farmworker Health and Safety Institute, n.d.)

The Institute has worked with over ten organizations, and it follows the train-the-trainer model so the organizations can keep the work moving in perpetuity after the training is complete. Their workshops and trainings include preparing community members to do research, training for health providers and handlers about pesticide poisoning and pesticide-related illness, and leadership development. While this coalition is not as active as it once was, its programs have continued both through its member organizations and new collaborations they have formed with other farmworker organizations. This work was pioneering, and set the standards for programs in the areas that they created, including diagnósticos, promoter and handler trainings, health provider trainings, focus groups, and trainings on organic farming and farmworker leadership development.

Migrant Clinicians Network

Next, the Migrant Clinicians Network (MCN), founded in 1984:

> Has several areas of programmatic focus, while remaining nimble and responsive to issues that arise in clinics throughout the country. At present, MCN has specific initiatives in the areas of occupational/environmental health, family violence prevention, and continuity of care/bridge case management for mobile patients. Other areas of expertise include immunization, infectious diseases, global health, chronic illness, and health promotion.
>
> (Migrant Clinicians Network, n.d.a)

While they serve all poor migrant workers, they provide services that are of particular help to farmworkers. Among their programs are initiatives in occupational and environmental health, cancer prevention and survivorship support, and expertise in global health and chronic illness. They have a special interest in farmworker health and safety, and, as such, have worked in partnership to develop resources on the pesticide-related health risks of working in agriculture. They provide resources and information pertinent to migrant and seasonal farmworker health, focusing on injury and illness, documentation, and health-care access with a specific focus on advancing health justice (Migrant Clinicians Network, n.d.b).

Broad-based organizations focused on farmworker rights

Farm Labor Organizing Committee, AFL-CIO

The Farm Labor Organizing Committee (FLOC), AFL-CIO is an organization founded in 1967 by Baldemar Valasquez, who was inspired by the work of Cesar Chavez and the United Farm Workers (Rosenbaum, 1993).[1] Focused on organizing farmworkers for change through union activity, it states, "FLOC has thousands of members between Ohio, North Carolina, South Carolina, and Mexico. Some of our members live in the United States, while other travel back and forth from their home country for the harvest season" (Farm Labor Organizing Committee, AFL-CIO, n.d.). They organize in the South, Mexico, and the Midwest, and have won victories in each of these areas. In addition, they organize community members to help support farmworkers in a number of different ways, including employment readiness and youth empowerment. All of their work is in support of their broader mission that we are all, "called upon to challenge the deplorable conditions of the broader workforce that remains voiceless, powerless, and invisible to mainstream America" (ibid.).

Earthjustice

Earthjustice is an environmental law organization that has many focus areas, including pesticides and toxic chemicals. On this front, they have been centrally involved in the updated Worker Protection Standard, and state of these changes, "Agricultural workers will now be better protected on many fronts – as long as the agencies responsible for implementing and enforcing the new rules do their job" (Earthjustice, 2015) and on this same webpage they state as their goal, "to connect policy set in our nation's capital with the reality of life on the frontlines of the agriculture system." On their page regarding their advocacy campaign to protect people from pesticides, they detail a number of initiatives they are involved in and statistics to raise awareness of the plight of farmworkers, all in service of their broader mission to use, "the power of law to defend our right to a healthy environment" (Earthjustice, 2014).

California Rural Legal Assistance Foundation

Like Earthjustice, the California Rural Legal Assistance Foundation is an organization that uses and provides legal resources to make social change. One of their many projects, known as the Agricultural Worker Health Project (AWHP), centers on agricultural worker health. The AWHP:

> Monitors and improves the health and safety of farm workers in the fields and in their homes. AWHP conducts hundreds of inspections each year and works with state agencies to enforce state laws designed to protect farm workers. AWHP goes one step further by monitoring state agencies

> to ensure that they are providing adequate protection to workers. Some of the issues which AWHP addresses are: pesticides, driverless tractors, unsafe or unhealthy work conditions, heat illness, lack of safety training or equipment, and sexual harassment.
>
> (CRLA, n.d.)

Given that California is a state in which agriculture is one of its predominant and major industries, changes there often lead the way for changes in other parts of the country. The work of the California Rural Legal Assistance Foundation not only helps individual farmworkers, but helps to lead the way on important policy-level change.

National Farm Worker Ministry

The National Farm Worker Ministry (NFWM) is:

> A faith-based organization that supports farm workers as they organize for justice and empowerment. NFWM began in 1920 as a ministry of charity and service, providing food, clothing and day care to the farm workers. When United Farm Workers founder Cesar Chavez began organizing in the 1960's, he called on the religious community to change its emphasis from charity to justice. NFWM became the vehicle for people of faith to respond to that call. NFWM brings together national denominations, state councils of churches, religious orders and congregations, and concerned individuals to act with the farm workers to achieve fundamental change in their living and working conditions. Grounded in faith, NFWM works side by side with farm workers throughout the country, organizing vigils, picketing, coordinating boycotts and educating constituents.
>
> (NFWM, 2012)

They assist with campaigns, and have a number of active farmworker organizations that serve as partners. Given its spiritual foundations, flexible approaches to change, involvement of multiple stakeholders, and focus on achieving just and equitable societal conditions, the NFWM was cited among other organizations as a resource in a chapter by Jennie McLaurin about migrant farmworkers. She states that, "cultural proficiency in population care is an ongoing task and assisted by acquisition of language skills, development of cross-cultural relationships, and global health experience. Advocacy for this marginalized immigrant labor force is an essential public health imperative" (McLaurin, 2014, p. 236). The focus of the NFWM on farmworker rights as they intersect with broader systemic policy issues are showcased in its many connections with other organizations and campaigns. One such campaign is the Sakuma Bros. Berry Boycott in Washington State. Called and organized by Familias Unidas por la Justicia (FUJ) in response to wage theft, racial harassment, and other labor injustices by Sakuma Bros., the NFWM endorsed the campaign and created multiple resources for support and awareness. Their commitment provides a network for information sharing, action alerts, solidarity coalitions, and additional resources to groups like FUJ seeking recognition from employers as to their representation of farmworkers for employment issues and justice (NFWM, 2016).

Pesticide Action Network North America

The Pesticide Action Network North America (PANNA) works:

> To replace the use of hazardous pesticides with ecologically sound and socially just alternatives. As one of five PAN Regional Centers worldwide, we link local and international consumer, labor, health, environment and agriculture groups into an international citizens' action network. This network challenges the global proliferation of pesticides, defends basic rights to health and environmental quality, and works to ensure the transition to a just and viable society.
>
> (PANNA, 2012)

This organization is actively involved in the creation of a domestic fair trade movement that shares aims with the movement that is already benefiting the global south. They support coalitions and help to network them with the idea of moving all members more quickly toward their goals. They also provide research and a great deal of power to member and partner initiatives. Perhaps most importantly, they use their resources as a large, international, and well-funded organization to help others achieve their goals and to keep pesticide use and its negative effects in the public eye. Again, while not primarily a justice-focused organization, PANNA represents models of justice based on collaboration, participation, and recognition of the unique strengths and voices everyone brings to the table.

Coalition of Immokalee Workers

In recent years, and since the movie release of *Food Chains*, directed by Sanjay Rawal, the Coalition of Immokalee Workers (CIW) has received a great deal of attention and recognition for its work with and for the farmworker community. The CIW is:

> A worker-based human rights organization internationally recognized for its achievements in the fields of social responsibility, human trafficking, and gender-based violence at work. Built on a foundation of farmworker community organizing starting in 1993, and reinforced with the creation of a national consumer network since 2000.
>
> (Coalition of Immokalee Workers, n.d.)

In this section, I will discuss their initiatives and programs emerging from that foundation. I would like to note here that the work of CIW differs in its tactics from many of the organizations already detailed in this chapter and book. It organizes farmworkers in campaigns aimed at some of the biggest food distributors and chains in the United States to raise awareness of the farmworker plight, but also to shift the amount of money going to farmworker wages in these high-volume markets. Andre Drainville comments on the symbolism of their work, asserting:

> In all their specificity, CIW actions present us with a concrete organizational experience from which we can think more generally about the political significance of what has been variously and vaguely termed 'the new internationalism of social movements', 'the anti-globalization movement', or 'globalization from below'. It also tells us a little about how global subjects might actually be present at their own making on the terrain of the world economy.
>
> (2008, p. 359)

CAMPAIGNS AND INITIATIVES

It is through the Fair Food Program (FFP) that CIW got its start around what they coined worker-driven social responsibility (WSR). It is through this program that CIW has successfully and famously pressured retail buyers to pay a penny more per pound for tomatoes, the results of which were channeled directly back to workers. This work has now extended to strawberries and peppers in Florida. They do this work through education of workers, compliance assurance, and standards enforcement created by:

> Market consequences guaranteed by CIW's legally binding Fair Food Agreements, in which participating buyers commit to buy Florida tomatoes only from growers in good standing with the FFP, and to cease purchases from growers who have failed to comply with the code of conduct.
>
> (Coalition of Immokalee Workers, n.d.)

They state that workers, growers, and retailers win out in the system because it not only increases protection of human rights for workers, but also increases effectiveness of risk management and increased positive reputation for growers and retailers associated with the program.

They also run an Anti-Slavery Campaign, focused on uncovering and investigating "multi-state farm slavery operations across the Southeastern United States, helping liberate over 1200 workers held against their will since the early 1990s" (ibid.). In addition, their model has been used in other areas in and outside of the United States and in industries beyond farming to fight against many types of forced labor. Finally, through its Campaign for Fair Food, the CIW "educates consumers on the issue of farm labor exploitation – its causes and solutions – and forges alliances between farmworkers and consumers in an effort to enlist the market power of major corporate buyers to help end that exploitation" (ibid.).

AFFINITY GROUPS

The Alliance for Fair Food is a coalition of people working in solidarity with the Coalition of Immokalee Workers around farmworker rights. They state:

> Following the tremendous leadership of the Coalition of Immokalee Workers, the ally network's tenacity, creativity, spirit and commitment have been integral to the unprecedented victories for workers that are changing the landscape of human rights in U.S. agriculture. Drawing from our collective strengths and successes, our mission remains to stand with farmworkers as they continue their struggle for justice and dignity.
>
> (Alliance for Fair Food, n.d.)

Currently, they are working on campaigns to pressure Wendy's and Publix (one of the largest fast food chains and one of the nation's leading grocery chains) to join the Fair Food Program.

Like the Alliance for Fair Food, the Student/Farmworker Alliance is a group of students and youth nationwide interested in changing our current system to ensure justice, respect, and dignity for farmworkers. They have three key principles in their organizing work: (1) working with—not for—farmworkers, (2) the tenet that students and youth have a powerful voice in working with farmworkers for systemic change, and (3) a focus on long-term movements for collective liberation (Student Farmworker Alliance, n.d.a). According to their website:

> It is a decentralized network, defined by the character and creativity of a multitude of communities and drawn together by our collective commitment to follow farmworker leadership. We build our network through education and action, and – in the spirit of CIW's "Consciousness + Commitment = Change" philosophy – we believe that both consciousness and commitment are a necessary part of a vibrant movement for justice.
>
> (Student Farmworker Alliance, n.d.b)

Together, these alliances and the programs that CIW has birthed hold a central place in many public consumers' thoughts and beliefs about farmworker justice and how to make change. While their approaches and tactics differ from many farmworker groups, their work has undoubtedly been influential in this sphere.

Certification and justice for workers

There is no shortage of labels meant to persuade consumers into buying products they feel best align with their values. However, many of these certification schemes, labels, and logos are light on what they mean for workers, and as covered previously, many consumers do not educate themselves about what these labels represent, particularly when they come to labor. Although many people, if pressed, would say that buying things with these labels likely positively impacts those who labor in the food system and on farms, they do this with little to no evidence of its truth. Furthermore, as Brown and Getz point out, certification plays into the neoliberal market system within which farmworkers are marginalized in the first place. They state:

> While the decline of agricultural unionism in California may suggest a failure of farm worker institutions and the collective bargaining model, the void created cannot be willed simply by market mechanisms. The social label is an inadequate substitute for the union label, yet its proliferation may foreclose the possibility of other forms emerging, which facilitate the collective action and agency of labor by prioritizing the agency of consumers.
>
> (2008a, p. 1194)

While I have previously covered organic certification, in this section I hope to highlight some of the work being done around certification and labeling particularly as a potential catalyst of change for farmworkers.

Domestic Fair Trade Association

The Domestic Fair Trade Association (DFTA) focuses on a number of different principles that fall under the broad categories of health, justice, and sustainability, including "farm worker rights and small farms in the United States" (Stenn, 2013, p. 10). In her book on fair trade and justice, Tamara Stenn notes regarding the definition of fair trade that:

> Mission statements among the principal Fair Trade institutions are similar and use terms such as "sustainable," "strengthen," "improve," "empower," and "promote." However, how the missions of sustainable empowerment and growth are achieved differs among institutions. Some institutions certify products, while others do not. Some recognize large-scale agriculture and hired labor as part of Fair Trade while others only recognize small farmers and their families. Fair Trade certifiers focus primarily on health, safety, labor, and pricing while noncertifiers are more focused on organizational business practices, networking, and business development. While not mutually exclusive, these focal differences help build diversity and expand justice within Fair Trade with plenty of controversy and debate.
>
> (ibid., p. 18)

The role of the DFTA, formally founded in 2007, states as its goals

> to support family-scale farming, to reinforce farmer-led initiatives such as farmer co-operatives, to ensure just conditions for agricultural workers, and to bring these groups together with mission-based traders, retailers and

> concerned consumers to contribute to the movement for sustainable agriculture.
>
> (DFTA, n.d.a)

It began working toward its goals by translating the principles of the international fair trade movement to the domestic market in the United States. In doing so, the organization hopes to help consumers, producers, growers, retailers, and farmworkers themselves navigate the tricky waters of meaning behind the multitude of fair trade labels that Stenn discusses. It is important to note that the "DFTA does not develop its own standards of domestic fair trade, but rather evaluates other standards, labels, and initiatives in the light of its own domestic fair trade principles, in order to provide information about domestic fair trade" (Kröger & Schäfer, 2013, p. 53).

One of their principles is the Rights of Labor, which states:

> Fair Trade means a safe and healthy working environment for producers and workers and conforms to all International Labour Organization conventions and the Universal Declaration of Human Rights. The participation of children (if any) does not adversely affect their well-being, security, educational requirements and need for play, and conforms to the United Nations Convention on the Rights of the Child as well as pertinent local/regional laws. Fair Trade ensures that there are mechanisms in place through which hired labor has an independent voice and is included in the benefits of trade through mechanisms such as living wages, profit sharing, and cooperative workplace structures. Apprenticeships are promoted to develop the skills of the next generation of farmers, artisans, and workers.
>
> (DFTA, n.d.a)

It is no surprise then that farmworker organizations are among the four different organizational membership categories in the DFTA, which are "(1) farmer organization, (2) farmworkers' organization, (3) retailer/processor/marketer/manufacturer and (4) non-governmental organization" (Duram & Mead, 2014, p. 156). Finally, as Sandy Brown and Christy Getz state, the DFTA "is the primary arena in which the meaning of domestic fair trade is being negotiated in the US" (Brown & Getz, 2008b, p. 12). Brown and Getz also point out that:

> It is worth noting that these members represent the "mission-driven" end of both the international fair trade and domestic alternative agro-food communities. As a result, they have consciously framed their work in opposition, not only to the conventional agrofood system, but also to industrialization and corporatization of both Fair Trade and organic certification systems.
>
> (ibid., p. 17)

In this same piece, the authors problematize the conflation of farmer and farmworker needs, but do recognize how in small-scale agriculture the livelihoods of these two groups are inextricably linked while also potentially at odds with one another.

One of the biggest projects undertaken by the DFTA has been the development of an extensive evaluation of various labels' fair trade and social justice claims along its developed criteria. A rubric was developed to assess how each label compared to the DFTA's criteria, with an ultimate assessment of the DFTA's ability to endorse the label for its fair trade and social justice implications (with a zero to five star rating scale). These evaluations happen in teams who review label documentation independently, then come together to discuss and come to a consensus about whether or not the label meets the DFTA criteria and at what level if they do. The evaluations are then reviewed for accuracy and a response from the certification body being evaluated. The published reviews currently assess Food Justice Certified (the label of the Agricultural Justice Project), Fair for Life (the certification program for the Institute for Marketecology), Fairtrade USA, Fairtrade International, Rainforest Alliance, Food Alliance, and the Equitable Food Initiative (DFTA, n.d.b). Only the Food Justice Certified and Equitable Food Initiative labels meet or exceed the DFTA guidelines for decent working conditions for farm and food workers, according to the main page of the fair facts evaluation website. Of all of the labels evaluated, only the Food Justice Certified label has received five stars, meaning it has met or exceeded the bar in all of the principal areas. The DFTA then provides a written summary of its evaluation of each label, including specific areas where it recommends improvement.

Agricultural Justice Project

The Food Justice Certified label, which received the highest rating from the DFTA is the labeling program of the Agricultural Justice Project (AJP), which is:

> A non-profit initiative in the U.S. which supports fairness in the food system through development of social justice standards for organic and sustainable agriculture, aiming at the creation of universal social standards based on the principles of International Federation of Organic Agriculture Movements (IFOAM), the Food and Agriculture Organization of the UN (FAO) and the International Labour Organization (ILO).
>
> (Kröger & Schäfer, 2013, p. 53)

The organization:

> Developed as a coalition effort among several organizations, including, the Rural Advancement Foundation International-USA (RAFI-USA), Peace Work Organic Farms, El Comité de Apoyo a los Trabajadores Agricolas (CATA), and Florida Organic Growers/Quality Certification Services (FOG/QCS). These groups were all working, albeit from different angles,

> to promote a holistic vision of sustainability via a comprehensive set of social justice standards.
>
> (Brown & Getz, 2008b, p. 17)

Brown and Getz acknowledge that these standards were meant to bridge the interests of family-scale farmers, buyers, and those of people laboring on the farms. AJP's label purposefully represents a high bar when it comes to labor, and as the authors acknowledge, this has served to push the DFTA to more carefully consider the dynamics associated with labor in the fair trade certification systems it evaluates. The founders of AJP are motivated by their chagrin at the lack of inclusion of labor in the organic certification standards of the USDA (Agricultural Justice Project, n.d.a). The Food Justice Certified label is unique in that it is democratically governed by workers, farmers, and businesses and the audit includes worker representatives who are trained by AJP (Agricultural Justice Project, n.d.b). In a recent AJP newsletter piece aimed at explaining to readers the rights of farmworkers, Elizabeth Henderson sums up the heart of the work of AJP when she states:

> Maybe the status of farmworker is a step towards owning and running a farm. Or perhaps farm work will continue for many years and may even be a person's chosen life work. Our movement for a local, sustainable agriculture needs both more farmers and more professional farm workers. Knowing your rights as a worker (or your responsibilities as an employer) provides the basis for a positive relationship so that small farms function as integrated teams.
>
> (Justice Feed, n.d.)

Equitable Food Initiative

The Equitable Food Initiative (EFI) is an organization that has another label evaluated by the DFTA. According to its website, EFI is:

> A unique partnership among businesses and organizations that have come together to develop standards, training processes and a certification to protect farm workers and produce safer, healthier food. This approach creates additional value and quality throughout the food system, benefiting workers, growers, retailers and consumers alike.
>
> (EFI, n.d.a)

They involve farmworkers among the stakeholders involved in overseeing and verifying certification, stating:

> Throughout the certification process, an EFI-trained "Leadership Team" helps the farm management and workforce to achieve compliance with the

> Standard, as confirmed by a third-party audit, and verify ongoing conformity, thereby reducing the likelihood of future labor violations, produce contamination, or pesticide hazards on the farm.
>
> (EFI, n.d.b)

The information published by the EFI appears, at least in theory, to reflect the interests of farmworkers, but there are some concerns reflected by the DFTA's evaluation of their label regarding what that process looks like on the ground.

Fair Food Program

As discussed previously, the Fair Food Program is an agreement initiative developed, supported, and overseen by the Coalition of Immokalee Workers. It is through this program that they enact their Worker-Driven Social Responsibility framework which is "premised on the participation of workers as the front-line monitors of their own rights in the workplace" (Fair Food Program, n.d.a). The agreements with participating buyers represent two provision areas: food premium and market enforcement. The food premium is a small increase in the price per unit of different fruits and vegetables, beginning with its campaign around tomatoes. This increase is to go directly back into the paychecks of farmworkers. The market enforcement provision requires participating buyers, "to suspend purchases from growers who have failed to comply with the Code of Conduct" (ibid.). These provisions, coupled with worker education about their rights, the development of complaint mechanisms for workers, and regularly conducted audits in which large numbers of workers participate, are meant to ensure the accountability of those participating in the program. The program boasts impressive results that have directly impacted the lives of thousands of workers (Fair Food Program, n.d.b).

Conclusion

In conclusion, one must ask about the impacts of these various channels of potential change on the issues facing farmworkers, as detailed in Chapter 2. It is especially important, considering what we have reviewed about justice and change throughout this book, to examine the various ways the organizations, initiatives, labels, and programs in this chapter truly recognize the power dynamics at play within farm labor, acknowledge the deeply rooted, historic dimensions of those injustices, and encourage participation and recognition of farmworkers in dignifying and better understanding the contributions of their work. They represent a variety of approaches to those intersections, and are held up to represent the diverse ways people around the United States engage in work to improve the plight of marginalized workers who are part of the subaltern of the American food system.

Note

1 See this piece by Rosenbaum for a detailed history of the organization.

References

Agricultural Justice Project. (n.d.a). Food Justice Certified. Available at: http://agriculturaljusticeproject.org/?page_id=13 (accessed July 19, 2015).

Agricultural Justice Project. (n.d.b). History: Food Justice Certified. Available at: http://agriculturaljusticeproject.org/?page_id=18 (accessed June 20, 2016).

Alliance for Fair Food. (n.d.). Our History. Available at: http://www.allianceforfairfood.org/our-history/ (accessed June 12, 2016).

Arcury, T. A., & Marín, A. J. (2009). Latino/Hispanic Farmworkers and Farm Work in the Eastern United States: The Context for Health, Safety, and Justice. In S. A. Quandt, & T. A. Arcury (Eds.), *Latino farmworkers in the Eastern United States* (pp. 15–36). New York Springer. Available at: www.springerlink.com/content/n796887132220894/abstract/

Bentz, V., & Shapiro, J. J. (1998). *Mindful inquiry in social research.* Thousand Oaks, CA: Sage.

Blackwell, M. (2006). *Líderes Campesinas: Grassroots Gendered Leadership, Community Organizing, and Pedagogies of Empowerment.* New York: NYU/Wagner Research Center for Leadership in Action. Available at: https://wagner.nyu.edu/files/leadership/Lideres.pdf

Blackwell, M. (2010). Lideres Campesinas: Nepantla Strategies and Grassroots Organizing at the Intersection of Gender and Globalization. *Aztlan: A Journal of Chicano Studies,* 35(1), 13–47.

Brown, S., & Getz, C. (2008a). Privatizing Farm Worker Justice: Regulating Labor Through Voluntary Certification and Labeling. *Geoforum,* 39(3), 1184–1196. Available at: http://doi.org/10.1016/j.geoforum.2007.01.002

Brown, S., & Getz, C. (2008b). Towards Domestic Fair Trade? Farm Labor, Food Localism, and the "Family Scale" Farm. *GeoJournal,* 73(1), 11–22. Available at: http://doi.org/10.1007/s10708-008-9192-2

Carrasquillo, N. (2011). Race and Ethnicity from the Point of View of Farm Workers in the Food System. *Race/Ethnicity: Multidisciplinary Global Contexts,* 5(1), 121–131. Available at: http://doi.org/10.2979/racethmulglocon.5.1.121

CATA. (2012). The Farmworker Support Committee Available at: www.cata-farmworkers.org

Chambers, R. (1997). *Whose reality counts?: Putting the first last.* London: Intermediate Technology.

Coalition of Immokalee Workers. (n.d.a). Campaign for Fair Food. Available at: www.ciw-online.org/campaign-for-fair-food/

Coalition of Immokalee Workers. (n.d.b). About CIW. Available at: www.ciw-online.org/about/

Coming Clean, Inc. (n.d.a). Farmworker Health and Justice Team. Available at: www.comingcleaninc.org/programs/farmworker-justice (accessed April 25, 2016).

Coming Clean, Inc. (n.d.b). Our History. Available at: www.comingcleaninc.org/about-us/history (accessed June 7, 2016).

Couto, R. (1999). *Making democracy work better: mediating structures, social capital, and the democratic prospect.* Chapel Hill, NC: University of North Carolina Press.

Couto, R. (2002). *To give their gifts: health, community, and democracy* (1st ed.). Nashville, TN: Vanderbilt University Press.

CRLA. (California Rural Legal Assistance, Inc.). (n.d.). Agricultural Worker Health Project Available at: www.crla.org/agricultural-worker-health-project (accessed June 12, 2016).

DFTA. (n.d.a). Evaluation of Fair Trade and Social Justice Certification Claims. Available at: http://fairfacts.thedfta.org/ (accessed June 20, 2016).

DFTA. (n.d.b). Vision, Mission, and Principles. Available at: www.thedfta.org/about/vision-mission-and-principles/ (accessed June 20, 2016).

Drainville, A. C. (2008). Present in the World Economy: The Coalition of Immokalee Workers (1996–2007). *Globalizations*, 5(3), 357–377. Available at: http://doi.org/10.1080/14747730802252495

Duram, L., & Mead, A. (2014). Exploring Linkages between Consumer Food Co-operatives and Domestic Fair Trade in the United States. *Renewable Agriculture and Food Systems*, 29(2), 151–160. Available at: http://doi.org/10.1017/S1742170513000033

Earthjustice. (2014, July 11).Protecting People from Pesticides. Available at: http://earthjustice.org/advocacy-campaigns/pesticides (accessed June 12, 2016).

Earthjustice. (2015, October 15). Harvesting Change: From Farms to the Nation's Capital. Available at: http://earthjustice.org/features/harvesting-change (accessed June 12, 2016).

EFI. (Equitable Food Initiative). (n.d.a). How You Can Participate. Available at: www.equitablefood.org/#!certification/c24gy (accessed June 20, 2016).

EFI. (Equitable Food Initiative). (n.d.b). Who We Are. Available at: www.equitablefood.org/#!executiveboard/c2i (accessed June 20, 2016).

Fair Food Program. (n.d.a). About the Fair Food Program. Available at: www.fairfoodprogram.org/about-the-fair-food-program/ (accessed June 20, 2016).

Fair Food Program. (n.d.b). Results. Available at: www.fairfoodprogram.org/results/ (accessed June 20, 2016).

Fals-Borda, O., & Anisur Rahman, M. (1991). *Action and knowledge: breaking the monopoly with participatory action research.* London: Apex Press.

Farm Labor Organizing Committee, AFL-CIO. (n.d.). Available at: www.floc.com/wordpress/cross-border-organizing/

Farmworker Health and Safety Institute. (n.d.). Available at: www.cata-farmworkers.org/english%20pages/fhsi.htm (accessed June 7, 2016).

Food Chain Workers Alliance. (2012). Food Workers Organizations in the U.S. Los Angeles, CA. Available at: http://foodchainworkers.org/wp-content/uploads/2012/11/2012-11-Food-Workers-Organizations-in-U.S.-FCWA.pdf

Freire, P. (2000). *Pedagogy of the oppressed* (30th anniversary ed.). New York: Continuum.

Green, G., & Haines, A. (2001). *Asset building and community development.* London: SAGE.

Green, J. J., Green, E. M., & Kleiner, A. M. (2011). From the Past to the Present: Agricultural Development and Black Farmers in the American South. In A. H. Alkon, & J. Agyeman, *Cultivating Food Justice: Race, Class, and Sustainability* (pp. 47–64). Cambridge, MA: MIT Press.

Greenwood, D., & Levin, M. (1998). *Introduction to action research: social research for social change.* Thousand Oaks, CA: Sage Publications.

Ife, J., & Tesoriero, Frank. (2002). *Community development: community-based alternatives in an age of globalisation* (2nd ed.). Frenchs Forest, NSW: Pearson Education.

Johnson, A. G. (2006). *Privilege, power, and difference.* Boston: McGraw-Hill.

Justice Feed. (n.d.). Newsletter. Available at: http://us8.campaign-archive2.com/?u=8e91198ca63649f1d81dc162f&id=af5252a56d&e=287171e3b1#knowrights (accessed June 20, 2016).

Kretzmann, J., & McKnight, J. P. (1996). Assets-based community development. *National Civic Review*, 85(4).

Kröger, M., & Schäfer, M. (2013). Between Ideals and Reality: Development and Implementation of Fairness Standards in the Organic Food Sector. *Journal of Agricultural and Environmental Ethics*, 27(1), 43–63. Available at: http://doi.org/10.1007/s10806-013-9444-0

La Via Campesina. (n.d.a). Food Justice and Food Sovereignty in USA. Available at: http://viacampesina.org/en/index.php/publications-mainmenu-30/1863-food-justice-and-food-sovereignty-in-usa (accessed June 12, 2016).

La Via Campesina. (n.d.b). Organisation. Available at: www.viacampesina.org/en/index.php/organisation-mainmenu-44 (accessed June 12, 2016).

Lideres Campesinas. (n.d.a). History. Available at: www.liderescampesinas.org/english/history.php (accessed June 7, 2016).

Lideres Campesinas. (n.d.b). Welcome to the Official Website. Available at: www.liderescampesinas.org/english/ (accessed June 7, 2016).

Migrant Clinicians Network. (n.d.a). The Migrant/Seasonal Farmworker Available at: www.migrantclinician.org/issues/migrant-info/migrant.html (accessed June 7, 2016).

Migrant Clinicians Network. (n.d.b). Who We Are. Available at: www.migrantclinician.org/about.html (accessed June 7, 2016).

Minkler, M., & Wallerstein, N. (2003). *Community-based participatory research for health.* San Francisco, CA: Jossey-Bass.

National Farm Worker Ministry. (2012, January). Together with Farm Workers Harvesting Justice. Available at: www.nfwm.org

National Farm Worker Ministry. (2016). Familias Unidas por la Justicia. Available at: http://nfwm.org/familiasunidasporlajusticia/ (accessed 23 June 2016).

Newell, P. (2005). Race, Class and the Global Politics of Environmental Inequality. *Global Environmental Politics*, 5(3), 70–94. Available at: http://doi.org/10.1162/1526380054794835

PANNA. (Pesticide Action Network North America). (2012). Available at: www.panna.org

Phillips, R., & Pittman, R. H. (2009). *An introduction to community development.* London: Taylor & Francis.

Rao, P., Arcury, T. A., & Quandt, S. A. (2004). Student Participation in Community-Based Participatory Research to Improve Migrant and Seasonal Farmworker Environmental Health: Issues for Success. *The Journal of Environmental Education*, 35(2), 3–15. Available at: http://doi.org/10.3200/JOEE.35.2.3-15

Rose, F. (2000). *Coalitions across the class divide: lessons from the labor, peace, and environmental movements.* Ithaca, NY: Cornell University Press.

Rosenbaum, R. P. (1993). Farm Labor Organizing Committee (FLOC): Grassroots Organizing for the Empowerment of the Migrant Farm Worker Community. *Culture & Agriculture*, 13(47), 21–23. http://doi.org/10.1525/cuag.1993.13.47.21

Rosset, P. (2011). Food Sovereignty and Alternative Paradigms to Confront Land Grabbing and the Food and Climate Crises. *Development*, 54(1), 21–30. Available at: http://doi.org/http://dx.doi.org.ezproxy.neu.edu/10.1057/dev.2010.102

Schlosberg, D. (2007). *Defining environmental justice: theories, movements, and nature.* New York: Oxford University Press.

Stenn, T. L. (2013). *The cultural and political intersection of fair trade and justice : managing a global industry*. New York: Palgrave Macmillan. Available at: http://site.ebrary.com/lib/alltitles/docDetail.action?docID=10800115

Stringer, E. (1999). *Action research* (2nd ed.). Thousand Oaks, CA: Sage Publications.

Student Action with Farmworkers. (2012). *About Us*. Available at: http://saf-unite.org/

Student Action with Farmworkers. (n.d.). History & Accomplishments. Available at: www.saf-unite.org/content/history-accomplishments (accessed June 7, 2016).

Student Farmworker Alliance. (n.d.a). Our Structure. Available at:, from http://www.sfalliance.org/our-structure/ (accessed June 12, 2016).

Student Farmworker Alliance. (n.d.b). How We Roll . Available at: www.sfalliance.org/how-we-roll/ (accessed June 12, 2016).

Toxic Free NC. (n.d.). About Us. Available at: http://www.toxicfreenc.org/about-us/ (accessed June 7, 2016).

UFW. (n.d.). The Official Web Page of the United Farm Workers of America. Available at: www.ufw.org/_page.php?menu=about&inc=about_vision.html (accessed June 6, 2016).

Wiggins, M. (2009). Farm Labor and the Struggle for Justice in the Eastern United States. In T. Arcury, & S. Quandt (Eds.), *Latino farmworkers in the Eastern United States* (pp. 201–220). New York: Springer.

Conclusion

What will it take to change the story?

Farmworkers in the United States are a community, and while they have been left out of many environmental justice (EJ) conversations, they are an important population to consider because of their unique challenges. Using an environmental justice framework not only allows us to understand farmworkers' issues in a more systemic way, but also illuminates more clearly leverage points for change. The environmental justice movement is rooted in creating a society that is productive and sustainable for all of its members, and it does this by focusing on how to make change with and for marginalized or subaltern populations who, through both systemic and on-the-ground injustice, bear the burden of pollution and other detrimental side effects of our decisions. Also, these are the very people who by definition are excluded from participating in the decision-making that impacts them.

Every day, people face multiple forms of oppression, domination, and exploitation. Farmworkers live their lives at the intersection of various marginalized identities and therefore experience multiple forms of oppression. Environmental justice and labor have a history of working together and being intricately linked, and therefore recognizing, valuing, and advocating for farmworkers as a community within the broader EJ movement have implications for changing the story – dignifying their work, creating channels through which they can fully participate in society and their own livelihoods, and eliminating the unjust conditions within which they often find themselves working and living. While the movement has focused on distributive justice and utilitarianism; productive, restorative, transformative, and transitional frames of justice may offer new ways of thinking about expanding the movement and increasing the efficacy of its approach and interventions. Perhaps most important is that one must have the freedom to choose as well as the potential to actively realize their capabilities for justice to be fully achieved. This book pulls together traditional thinking on justice and environmental justice, as well as new, innovative scholarship in the movement to put forward a model for looking at farm labor across the continuum of agriculture- from small- and family-scale organic agriculture to large-scale conventional as seen again in Figure 0.1 in the Introduction on p. 3.

The environmental justice movement was solidified at the summit in the 1990s, and it was at this time that the environment was named as a place where people live, work, and play and it was noted that there were disparities drawn on different lines of marginalization in how people access a healthy environment that allows them to live just, full lives. Farmworkers helped to actively shape the movement, and environmental justice also drew inspiration from early farmworker organizing. Yet, there are still problems when trying to advocate for farmworkers as a community within the broader environmental justice movement. In a personal email in 2016 from Jeannie Economos at the Farmworker Association of Florida, she summed it up by stating:

> It has been a challenge to connect farmworker environmental justice with the overall environmental justice movement. Farmworkers do not follow the patterns that are indicative of the majority of environmental justice communities that are more place-/location-based, such as a community that is next to an area of refineries and processing plants, etc. Farmworkers are a community that spans the country. Except for residing largely in rural, agricultural communities across the country, they are not place-based, and can be migrant and/or seasonal workers, and in some cases, are temporary workers coming to work for only a short time in the US, such as H-2A or guestworkers. This can make building community among some farmworker groups difficult (although the recent trend is for less and less migration among farmworkers along traditional migration routes). Part of the dynamics of the farmworker reality is by design to keep the agricultural workforce unorganized and disempowered.

She then explains that there is hope:

> That being said, I think one area that is ripe for bringing farmworker justice and environmental justice groups together is around children's health related to exposures to, for example, various kinds of endocrine-disrupting chemicals and other kinds of chemicals. There is also a growing interest among more "traditional" EJ groups in food security, food safety and access and food justice. Urban community gardens and increasing interest in creating spaces for EJ communities to have local, safe, fresh, healthy, pesticide-free food for their communities is one area that can begin to tie farmworkers and their reality to that of EJ communities. This needs to be more deeply explored and developed, but it is an area where EJ and FJ can find common interest, common ground, and, hopefully, around which to create common campaigns.

We must start with who works in agriculture, what injustices they face and why. As we explored in Chapter 2, there are migrant workers, seasonal workers, and guestworkers laboring on farms in the United States, and more than 50 percent of all the 2.7 million+ workers are not legally authorized to work in the United States. Additionally, hired laborers work mostly on large farms ($500,000 of sales per year or more) and there are more than 2 million unpaid workers. The average age of farmworkers is in the mid-thirties, and 70 percent of farmworkers report speaking English only somewhat to not at all.

When faced with making a profit and ensuring crops are harvested, the agricultural industry and its farm owners are able either to create just working conditions to attract more workers to its difficult, temporary jobs or to find and exploit people who really have no other options. The majority of owners have traditionally gone the second way, which has resulted in paternalistic relationships in which the workers are reliant upon employers not only for work but for their other basic needs as well. This happens largely because within the broader political economic landscape within which agriculture operates, the reward for farm owners associated with violating environmental and labor regulations is greater than the potential risk if they are discovered. Unsurprisingly, these dynamics create a context within which marginalization and injustice can easily occur. These injustices include occupational injury and illness (including pesticide-related illness), lack of access to healthcare, housing and hunger, issues pertaining to immigrant and migrant workers, issues unique to women and children working in agriculture, low wages, lack of stable employment, and the inability to organize. These injustices are not mutually exclusive, and most often relate to and compound one another.

Most think that the picture must be rosier when it comes to organic farms. However, even if it were, organic farms comprise less than 1 percent of the nation's total farms, and only 60 percent of those use hired labor. Additionally, justice issues are not taken into account in organic certification as defined and regulated by the United States Department of Agriculture (USDA), resulting in a myth that organic agriculture in itself is ethically and morally just. Organic agriculture is also more labor-intensive than conventional farming techniques, making it difficult to recruit and retain workers, and within our larger system of agriculture it is also difficult to make an income due to low sales and market potential and lack of market reach, particularly on very small- and family-scale farms. Because of these dynamics, while organic agriculture is tied up with larger ideals of reforming society through reforming the food system, it often falls into the same traps as agriculture at other scales.

In exploring how this came to be, one must look to history and policy to understand the conditions within which farmworkers face injustice, and it is also important to note that the lack of voice among farmworkers has become even more of an issue as policies and trade have increased the numbers of immigrants working in agriculture both historically and currently. Much of this is summed up in the literature as agricultural exceptionalism, or the exclusion of farmworkers from the legal protections that other workers have. Chapter 3 identifies the key policy areas for making changes for farmworkers as: labor, housing, pesticide, healthcare, and immigration policy. As shown in regard to labor relations and standards, there are restrictions and obstacles to organizing, specifically around minimum wage rates, overtime provisions, and child labor laws. In addition, policy more often than not serves to enforce and reinforce farmworkers' place as a marginalized (subaltern) population, stuck in a subordinated position to the benefit of the overall capitalist, neoliberal system of agriculture. There are particular shortcomings in laws and policies when it comes to women and children farmworkers.

When it comes to health, safety, and worker protection, the Occupational Safety and Health Administration (OSHA) exempts businesses with less than 11 employees from many of its regulations, meaning the majority of U.S. farms do not have to comply. In addition, enforcement, accountability, and farmworker education regarding policy are difficult due to variations from state to state. Furthermore, while the Environmental Protection Agency (EPA) recently updated the Worker Protection Standard (WPS), arguably the most important regulatory piece for pesticide use, information, and safety, questions remain regarding implementation and enforcement both due to obstacles of the past, as well as funding-related pressures.

Shortages during WWII led to the Bracero program (1942–1964) to bring in laborers from Mexico, which increased the number of workers in U.S. agriculture overall. This reduced farmworkers' ability to organize and bargain due to an oversupply of workers once the war was over. This dynamic set the stage for later guestworker policies and the H-2A program, and even as these immigration policies have made it easier to gain access to citizenship (such as through the Immigration Reform Control Act, or IRCA), they have also resulted in discrimination against immigrant employees. This, paired with trade policy that displaces peasant and subsistence farmers in other countries, sets up conditions within which those countries' economies rely on the export not only of goods, but also of labor. In some areas, remittances count for up to 50 percent of the local economy, ensuring that there is a steady supply of cheap labor with little to no bargaining power or paths for advocacy to deal with encountered injustices.

In Chapters 4 and 5, we saw how these dynamics play out on the ground in two case studies, one situated in conventional agriculture and the other in organic. Chapter 4 detailed the case of the Farmworker Association of Florida (FWAF), an organization formed in the early 1980s by farmworkers, and that works across the whole state through its five chapters. This case focused on the Apopka, Florida, community. The chapter of FWAF located there deals with both issues faced by people who have worked the land there for generations, as well as those faced by immigrant and migrant farmworker communities. In the 1990s when it became clear that there was an environmental catastrophe caused by the pesticides used on the farms around Lake Apopka, environmental, rather than human, concerns were preeminent for those agencies in power. Therefore, many farmworkers found themselves without work or housing and also suffering from illnesses caused by the same pesticides that had caused the environmental harm. The FWAF focuses on community building, healing, education, and participation through activities like pesticide training, participatory agroecology and community gardens, and its Memorial Quilt Project. In addition, the organization and its staff and members engage in ongoing advocacy on behalf of systemic policy change that will impact the overall context within which farmworkers labor. It does all of this while also focusing on the on-the-ground support needed from its membership community, and participating in and collaborating on research projects needed to advance policy positions. The FWAF has worked and still works in partnership with other organizations around the country and the world for a more just world for farmworkers.

As the study presented in Chapter 5 illustrates, farmers' and farmworkers' well-being are inextricably linked in small-scale organic agriculture. Although certain types of pesticides are permissible within organic labeling, and chemical drift and contamination from nearby farms can also occur, substandard housing, lack of benefits and health insurance, low pay, and health issues due to the physical nature of the work all remain primary concerns for organic farming. As this case study, conducted in partnership with the Northeast Organic Farming Association (NOFA) illustrates, there are concerns over the ability to create the context for justice in small- and family-scale organic agriculture. In addition, farmers and farmworkers in organic agriculture are driven more by an alignment of their values with this type of farming rather than the ability to earn a living or make a livelihood out of working in this type of agriculture. Study participants' conceptualizations of justice tied into concerns about ecological and environmental well-being (in addition to labor), education, and values rooted in community. In contrast to the case of FWAF, NOFA is a decentralized multi-state network that is mostly focused on advancing organic agricultural techniques and a philosophy, among which labor is only one dimension. In this study it became clear that it is difficult for such a network to leverage effectively for policy and other platforms that would benefit workers within the full network. Finally, this case reinforced ideas that organic agriculture is marginalized within the larger political economic system of agriculture, and most consumers conflate organic with more rights for workers.

Chapter 6 examined multiple pathways to change for farmworkers that are underway within agriculture and the broader environmental justice movement. Farmworkers are at the bottom of the chain of food production, and as long as people continue to want cheap food and other parts of that chain are not squeezed, the fight for farmworker justice is a difficult one. That said, more people do care about food now than ever before, access to information is easier, and we may be at a moment in time when the story can change. Many people are engaged in direct and indirect ways to dignify the work of farmworkers, to advocate for policy and process changes that will positively impact farmworkers' livelihoods and health, and to answer those fundamental questions of philosophies of justice on how they can live their best lives and have the agency to change their own conditions by having a place that is valued in our societal, economic, and political structures. Central to a discussion about the systems of power and privilege that have served as a backdrop for farmworker injustice is the recognition of the reinforcement of structures within which power exists, ensuring that those with privileged positions maintain them. Only with this recognition and drawing from models of change offered by the literature on both coalition building and asset-based community development, can we begin to think about change that truly recognizes the capabilities of the farmworker community and engages them in participatory change at the local *and* system levels. The examples in this chapter point to organizations and coalitions that are working in partnership with the farmworker community (or have emerged from that community itself) to create positive change by way of accessing resources, providing education about workplace hazards and protective measures, and policy change to improve the broader system. In addition, the chapter also illustrates ways that consumers can educate themselves to use their purchasing power for food that is not only sustainable, but also just.

I hope that recognizing farmworker justice within an environmental justice frame can raise these issues to a more prominent national scale and bring them to the attention of more people, and that doing so can also provide new ways of thinking about making change with and for our farmworker community. Thinking back on the day when I rode my bike through the very fields that put these workers, so essential to our very vitality, on my radar screen, I realize that I wish not to see a different scene per se, but to see one where, when the layers are pulled back, the work of these individuals is not only dignified but is also work they can identify with pride. I hope for a world where the injustices that farmworkers face are not supported by our very laws and policies, where agricultural work is *not* an exception, and one in which the larger political economic landscape allows for people to farm, live, work, and eat in alignment with their values and in fulfillment of their own existence. While this book paints a troubling and dire picture of farm work in the United States and the conditions that relegate farmworkers to being the subaltern, it also offers a promising new way for people advocating and working tirelessly for change in this arena to frame their efforts.

Addendum

As you know, moving a research project of this sort from conceptualization to production does not happen overnight. I completed this book in late summer of 2016, and a great deal has happened in the United States since then (at the time of this writing in late November 2016). Donald Trump has been elected President, and the Republican Party is now in control of both the House and the Senate. This election cycle was unprecedented in many ways, in its focus on character issues (and flaws) rather than substantive public policy positions. All progressive movements in the United States are currently bracing themselves for what is to come. Many of us are wondering what changes are in store for the country and the world given the explicit racism, xenophobia, homophobia, and misogyny expressed by the Trump forces during the election and afterwards. Perhaps we knew that these ugly sentiments were there all along among so many Americans. But like the sediment at the bottom of a riverbed being stirred up by a storm, this election brought it to the surface in a way that makes it more visible than ever before. I have concerns about this new administration's potential for thwarting the environmental justice (EJ) movement, and specifically the issues discussed throughout this book that impact farmworkers directly. Farmworkers are a subaltern population who experience multiple systems of oppression. As a result, government policies around housing, labor, pesticides, healthcare, and immigration all have direct and powerful consequences on their quality of life.

Using history as a guide, we can see that under previous corporate-friendly, conservative administrations the political platform expressed in the campaign is anti-immigrant. Once in office, however, such governments actually loosen regulations restricting immigrants from entering the country. This is likely due to several reasons, but most directly to the fact that a neoliberal, capitalist system relies on immigrant workers as a highly exploitable labor force, and there are few sectors in the economy where this is seen more clearly than in agriculture. As a result, while loosening immigration regulations or 'looking the other way' on their enforcement, these administrations engage in actions designed to minimize the political-economic power of immigrant labor, such as blocking paths to citizenship, denying collective bargaining and other labor rights, opposing occupational health and safety measures, and so on. Trump has come out fairly actively against the Trans Pacific Partnership (TPP) and the Transatlantic Trade and Investment Partnership (T-TIP), but he is for neoliberal trade agreements in general. It is unclear at this time what the fate of these two partnership agreements may be and/or what new negotiations may affect the flow of goods, services, and human capital across borders. In addition, calls for mass deportations and the reversal of DACA (Deferred Action for Childhood Arrivals), passed by President Obama through Executive Action and which allows a legal status and a work permit for people who entered the United States illegally prior to the age of 16, is causing fear among the nation's farmworkers. In a statement issued by the United Farm Workers (UFW), they state:

> One of the great ironies in the election of Donald Trump as the next president of the United States is that the agricultural heartland of the country voted overwhelmingly for a candidate who pledged to deport millions of undocumented workers, a move that would undermine the agricultural workforce and ripple out in the food economy in unforeseen, but likely negative, ways.

As outlined in Chapter 3, the 'designated representative' portion of the implementation phase of the 2015 revisions to the Worker Protection Standard (WPS) was already tenuous at best at the federal level due to conflicting support from Congress (arguably more representative of the interests of business/farm owners than of farmworkers), which ultimately funds the Environmental Protection Agency (EPA) who will oversee and direct funding for implementation efforts at the state level. The troubling early appointment of Myron Ebell to head the transition of the EPA represents a reconsolidation of power by the polluter-industrial complex over the state (Faber 2008). Ebell spells sure disaster on a number of issues, and his stated focus of dismantling efforts related to curbing climate change and regulate toxic chemicals ensures that the EPA is going to be crippled. It is also likely that the Trump administration will work to obstruct and dismantle the WPS, and even if they don't dismantle it completely, at least its implementation could be delayed indefinitely.

All of this amounts to a hard blow when it comes to justice for farmworkers, but it is my hope that these words by Arturo Rodriguez (UFW President since Cesar Chavez's death in 1993), emailed to Modern Farmer on the day after the election, ring true. He said:

> Making progress will be more difficult, but our practice of organizing for the long term will let us continue to win justice...We reaffirm our confidence in the American people and in our Si Se Puede! attitude...As the challenges increase so does our commitment and determination to afford all farm workers and low-income working families the opportunities and respect they deserve.

The work will continue, and must involve both analyzing and critiquing present systems of racial oppression and class exploitation while at the same time exploring emancipatory goals that allow the subaltern to move beyond their current condition. This quote (as seen in Chapter 6) by Robert Chambers perhaps sums up best what we have seen from Donald Trump thus far:

> The upper rejects discordant feedback. He re-labels the lower and redefines her reality for her. He seeks to transfer his reality. He wants it to be his reality that counts. What prevents him from learning in this case is not professionalism or bureaucracy. It is his dominant behavior, person to person. It is his power.

His desire to maintain power and manipulate the rhetoric through which people view their own conditions is perhaps his most dangerous trait of all, and those who advocate for justice for all people and our environment would do well to call this out as often as possible so that the realities of our most oppressed, marginalized peoples do not get defined for them, and, ultimately, for all of us.

Another event that has happened in the time that has transpired since the book went into production was the passing of Geraldean Matthew, mentioned and cited frequently throughout Chapter 4 for her work with the Farmworker Association of Florida.[1] I, along with many people, mourn her passing and use it as fuel for inspiration for what lay ahead at this time. Despite everything that had happened to Geraldean, and the multitude of health problems that stemmed from her repeated exposure to pesticides, she was an advocate for farmworkers. She would spend her days in ways few of us could imagine in her condition, such as going to a dialysis treatment in the morning, giving public testimony that same afternoon, then spending time helping farmworkers navigate access to healthcare and housing in the evening. She has shown us the way forward, and for that in these times I re-dedicate this work to her, and so many countless other heroes engaged in the struggle for social and environmental justice.

Note

1 You can read more about Geraldean's life and passing in this guest blog post, written by Jeannie Economos: http://earthjustice.org/blog/2016-september/in-honor-and-recognition-of-geraldean-matthew#

List of abbreviations

AFL-CIO	American Federation of Labor and Congress of Industrial Organizations
AgJOBS	Agricultural Job Opportunities, Benefits and Security Act
AJP	Agricultural Justice Project
ALRA	Agricultural Labor Relations Act
ARMS	Agricultural Resource Management Survey
AWHP	Agricultural Worker Health Project
BAWP	Border Agricultural Workers Project
CAFTA	Central American Free Trade Agreement
CATA	El Comite de Apoyo a Los Trabajadores Agricolas, Farmworker Support Committee
CBPM	Community-Based Prevention Marketing
CDC	Centers for Disease Control
CIW	Coalition of Immokalee Workers
CSA	Community Supported Agriculture
DFTA	Domestic Fair Trade Association
DOL	Department of Labor
EEOC	Equal Employment Opportunity Commission
EFI	Equitable Food Initiative
EJ	Environmental Justice
EPA	Environmental Protection Agency
ERS	Economic Research Service
FFP	Fair Food Program
FHJ	Farmworker Health and Justice
FHSI	Farmworker Health and Safety Institute
FIFRA	Federal Insecticide, Fungicide, and Rodenticide Act
FLOC	Farm Labor Organizing Committee
FLS	Farm Labor Survey
FLSA	Fair Labor Standards Act
FOG	Florida Organic Growers
FTE	full-time equivalent
FUJ	Familias Unidas por la Justicia
FUTA	Federal Unemployment Tax Act

FWAF	Farmworker Association of Florida
GDP	Gross Domestic Product
IFOAM	International Foundation for Organic Agriculture Movements
ILO	International Labour Organization
MCN	Migrant Clinicians Network
MOFGA	Maine Organic Farmers and Gardeners Association
MSPA	Migrant and Seasonal Worker Protection Act
NAFTA	North American Free Trade Agreement
NASDA	National Association of State Department of Agriculture
NASS	National Agricultural Statistics Service
NAWS	National Agricultural Workers Survey
NFWM	National Farm Worker Ministry
NLRA	National Labor Relations Act
NOFA	Northeast Organic Farming Association
PANNA	Pesticide Action Network North America
QCS	Quality Certification Services
RAFI-USA	Rural Advancement Foundation International-USA
SAF	Student Action with Farmworkers
SAW	Seasonal Agricultural Worker
TPP	Trans-Pacific Partnership
T-TIP	Transatlantic Trade and Investment Partnership
UFW	United Farm Workers
UN FAO	United Nations Food and Agriculture Organization
USDA	United States Department of Agriculture
WPS	Worker Protection Standard
WSR	Worker-Driven Social Responsibility

Index

Figures are shown by a reference in *italics* and tables are referenced in **bold**.

For Product Safety Concerns and Information please contact our EU
representative GPSR@taylorandfrancis.com
Taylor & Francis Verlag GmbH, Kaufingerstraße 24, 80331 München, Germany

www.ingramcontent.com/pod-product-compliance
Ingram Content Group UK Ltd.
Pitfield, Milton Keynes, MK11 3LW, UK
UKHW011011180926
475701UK00004B/680

* 9 7 8 0 3 6 7 2 2 7 7 9 1 *